THIRD INTERNATIONAL SYMPOSIUM ON AEROBIOLOGY

Supported by financial contributions from:

Ministry of Defence
Medical Research Council
Evans Medical Ltd.
The Pfizer Group
Imperial Chemical Industries Ltd.

ORGANIZING COMMITTEE

C. E. Gordon Smith *Chairman*
D. A. J. Tyrrell
R. E. O. Williams
I. H. Silver *Secretary*

CONVENERS

A. C. Allison
W. Brumfitt
H. M. Darlow
K. R. May
M. T. Parker
F. T. Perkins
I. H. Silver
C. E. Gordon Smith
D. A. J. Tyrrell
R. E. O. Williams

AEROBIOLOGY

Proceedings of the Third International Symposium held at the University of Sussex, England, September 1969

Edited by

I. H. SILVER

Microbiological Research Establishment, Porton Down Wiltshire, England

1970

ACADEMIC PRESS · LONDON · NEW YORK

ACADEMIC PRESS INC. (LONDON) LTD.
Berkeley Square House
Berkeley Square
London, W1X 6BA

U.S. Edition published by
ACADEMIC PRESS INC.
111 Fifth Avenue
New York, N.Y. 10003

Library of Congress Catalog Card Number: 72-129794
ISBN 0-12-643550-2

PRINTED IN GREAT BRITAIN BY
BUTLER & TANNER LTD
FROME AND LONDON

List of participants

AKERS T. G. Cdr. — *Naval Biological Laboratory, Naval Supply Center, Oakland, California 94625, U.S.A.*

ALLISON A. C. Dr. — *Clinical Research Centre Laboratories, N.I.M.R., Mill Hill, London N.W.7*

APPLEBY E. C. Dr. — *Royal Veterinary College, Royal College Street, London N.W.1*

ARTENSTEIN M. S. Dr. — *Department of Bacteriology, Walter Reed Army Institute of Research, Walter Reed Army Medical Center, Washington, D.C. 20012, U.S.A.*

ASH C. G. Dr. — *Deseret Test Center, Building 103, Soldiers Circle, Fort Douglas, Utah 84113, U.S.A.*

AYLIFFE G. A. J. Dr. — *Hospital Infection Research Laboratory, Summerfield Hospital, Western Road, Birmingham 18*

BARTLEMA H. C. Dr. — *Medisch Biologisch Laboratorium RVO/TNO, Lange Kleiweg 139, Rijswijk Z.H., Netherlands*

BEALE A. J. Dr. — *The Wellcome Research Laboratories, Langley Court, Beckenham, Kent*

BENBOUGH J. Dr. — *Microbiological Research Establishment, Porton, Salisbury, Wiltshire*

BETHEA W. R. Dr. — *Suite 103, Beverley Medical Building, 415 North Camden Drive, Beverley Hills, California*

BLOWERS R. Prof. — *Public Health Laboratory, Ayresome Green Lane, Middlesbrough*

BRACHMAN P. S. Dr. — *Bacterial Diseases Section, Epidemiology Program, National Communicable Disease Center, Atlanta, Georgia 30333, U.S.A.*

BUCKNALL R. A. Dr. — *Virus Research Department, I.C.I. Ltd., Pharmaceutical Division, Mereside, Alderley Park, Macclesfield, Cheshire*

BURKE J. Prof. — *Massachusetts General Hospital, Boston, Mass., U.S.A.*

BRUMFITT W. Dr. — *Department of Pathology, Edgware General Hospital, Edgware, Middlesex*

CHADWICK P. Prof. — *Department of Microbiology, Queen's University, Kingston, Ontario, Canada*

CHANOCK R. M. Dr. — *Laboratory of Infectious Diseases, National Institute of Allergies and Infectious Diseases, National Institute of Health, Bethesda, Maryland, U.S.A.*

CHARNLEY J. Dr. — *Regional Centre for Hip Surgery, Wrightington Hospital, Appley Bridge, Nr. Wigan, Lancs.*

COX C. S. Dr.	*Biophysical Research Branch, Physical Sciences Division, Fort Detrick, Frederick, Maryland 21701, U.S.A.*
DARBYSHIRE J. H. Dr.	*Animal Virus Research Institute, Pirbright, Woking, Surrey*
DARK F. A. Mr.	*Microbiological Research Establishment, Porton, Salisbury, Wiltshire*
DARLOW H. M. Dr.	*Microbiological Research Establishment, Porton, Salisbury, Wiltshire*
DAVIDS D. E. Mr.	*Bacteriology Section, Defence Research Establishment Suffield, Ralston, Alberta, Canada*
DAVIES R. R. Dr.	*Department of Bacteriology, Wright-Fleming Institute, St. Mary's Hospital Medical School, Paddington, London W.2*
DAWSON P. S. Dr.	*Central Veterinary Laboratory, New Haw, Weybridge, Surrey*
DONALDSON A. I. Dr.	*c/o Animal Virus Research Institute, Pirbright, Surrey*
DOWNIE J. Miss	*University of Sheffield, Department of Medicine, The Royal Hospital, Sheffield*
DRUETT H. A. Mr.	*Microbiological Research Establishment, Porton, Salisbury, Wiltshire*
DUBOVI E. Lt.	*Naval Biological Laboratory, Naval Supply Center, Oakland, California 94625, U.S.A.*
EHRLICH R. Dr.	*Life Sciences Research Department, I.I.T. Research Institute, Chicago, Ill. 60616, U.S.A.*
GARD S. Dr.	*Department of Virus Research, Karolinska Institute, Box 764, Stockholm, Sweden*
GARDNER P. S. Dr.	*Department of Virology, The Royal Victoria Infirmary, Queen Victoria Road, Newcastle upon Tyne 1*
GOLDBERG L. Dr.	*Naval Biological Laboratory, Naval Supply Center, Oakland, California, 94625, U.S.A.*
GORDON F. B. Dr.	*Department of Bacteriology, Naval Medical Research Institute, Bethesda, Maryland 20014, U.S.A.*
GOULD J. C. Dr.	*Central Microbiological Laboratories, Western General Hospital, Crewe Road, Edinburgh*
GRIST N. R. Prof.	*University of Glasgow, Department of Infectious Diseases, Regional Virus Laboratory, Ruchill Hospital, Glasgow N.W.*
HAMBLETON P. Dr.	*Microbiological Research Establishment, Porton, Salisbury, Wiltshire*
HARPER G. J. Mr.	*Microbiological Research Establishment, Porton, Salisbury, Wiltshire*
HAUSLER W. J. Dr.	*University of Iowa, State Hygiene Laboratory, Medical Laboratory Building, Iowa City, Iowa 52240, U.S.A.*
HELLMAN A. Dr.	*Biohazards Control and Containment Section, Etiology Area, National Cancer Institute, N.I.H., Bethesda, Maryland 20014, U.S.A.*
HERS J. F. Ph. Dr.	*Department of Clinical Respiratory Virology, University Hospital, Leiden, Netherlands*
HILDICK-SMITH G. Dr.	*Clinical Research Division, Johnson & Johnson Ltd., New Brunswick, New Jersey, U.S.A.*

HIRST J. M. Dr.	*Rothamsted Experimental Station, Harpenden, Herts.*
HOBSON D. Dr.	*Department of Bacteriology, The University, P.O. Box 147, Liverpool 3*
HOOD A. M. Mr.	*Microbiological Research Establishment, Porton, Salisbury, Wiltshire*
JAEGER R. F. Dr.	*Biological Consultant, Physical and Life Sciences Section, U.S.A.M.C.D., Scientific and Technical Information Team—Europe, 1–9 Hochhause Bremerstrasse, 6, Frankfurt am Main, Fed. Republic West Germany*
JAMESON B. Dr.	*The Royal Marsden Hospital, Fulham Road, London S.W.3*
JEWETT A. C. Mr.	*Microbiology Program, Office of Naval Research, Washington, D.C. 20306, U.S.A.*
de JONG J. C. Dr.	*Laboratorium voor Microbiologie, de Rijksuniversiteit, Cartharijnesingel 59, Utrecht, Netherlands*
KNIGHT V. Dr.	*Baylor University College of Medicine, Texas Medical Center, Houston, Texas 77025, U.S.A.*
KOHN A. Dr.	*Israel Institute of Biological Research, Ness-Ziona, P.O. Box 19, Israel*
LAMSON T. H. Col.	*Preventive Medicine Officer, H.Q. U.S.A.R.E.U.R., Office of the Surgeon General, A.P.O., New York 09403*
LARSON E. W. Mr.	*Applied Aerobiology Division, Fort Detrick, Frederick, Maryland 21701, U.S.A.*
LAURELL G. Dr.	*Institute of Medical Microbiology, Department of Clinical Bacteriology, University of Uppsala, Uppsala, Sweden*
LAWTHER P. J. Prof.	*Air Pollution Research Unit, St. Bartholomew's Hospital Medical College, Charterhouse Square, London, E.C.1*
LIDWELL O. M. Dr.	*Central Public Health Laboratory, Colindale Ave., London N.W.9*
LIEM K. S. Dr.	*Clinical Research Department, Philips-Duphar, Postbus 2, Weesp, Netherlands*
LINDBOM G. Dr.	*Institionen for Medicininsk Mikrobiologi, Avdelningen for Klinisk, Bacteriologi, Uppsala Universitet, Uppsala, Sweden*
LOVEDAY D. Dr.	*Fisons Pharmaceuticals Ltd., Bakewell Road, Loughborough, Leics*
LOWBURY E. J. Dr.	*Burns Unit, Birmingham Accident Hospital, Bath Row, Birmingham 15*
LUNDIN J. Dr.	*Research Institute of Swedish National Defence, Sundyberg 4, Sweden*
McHUGH D. P. Mr.	*Vaccine Division, the Pfizer Group, Sandwich, Kent*
McLEOD R. A. Dr.	*Department of Microbiology, MacDonald College of Magill University, P.Q., Canada*
MARPLES R. R. Dr.	*Department of Dermatology, University of Pennsylvania, Philadelphia, Pa. 19104, U.S.A.*
MAY K. R. Mr.	*Microbiological Research Establishment, Porton, Salisbury, Wiltshire*

MAZZARELLA M. A. Dr.	*Naval Submarine Medical Center, Box 60, Naval Submarine Base—New London, Groton, Connecticut 06340, U.S.A.*
MEINDERSMA T. E. Dr.	*Isolatie Paviljoen, Academisch Zeikenhuis, Leiden, Netherlands*
METZGER J. F. Col.	*Pathology Division, U.S. Army Medical Unit, Fort Detrick, Frederick, Maryland 21701, U.S.A.*
MILLER W. S. Dr.	*Becton-Dickinson Research Center, PO Box 11276, Raleigh, N. Carolina 27604, U.S.A.*
MILES A. Sir	*Lister Institute of Preventive Medicine, Chelsea Bridge Road, London S.W.1*
MORTIMER E. A. Dr.	*University of New Mexico, School of Medicine, Albuquerque, New Mexico 87106, U.S.A.*
MORTON J. D. Mr.	*Melpar Inc., 7700 Arlington Boulevard, Falls Church, Virginia 22046, U.S.A.*
MIMS C. Dr.	*The Wistar Institute, Thirty-sixth Street, at Spruce, Philadelphia, Pa. 19104, U.S.A.*
NASH T. Mr.	*Microbiological Research Establishment, Porton, Salisbury, Wiltshire*
NOBLE W. C. Dr.	*Department of Bacteriology, St. John's Hospital for Diseases of the Skin, Lisle Street, London W.C.2*
NORRIS K. P. Dr.	*Microbiological Research Establishment, Porton, Salisbury, Wiltshire*
NYSTROM B. Dr.	*Laboratory for Clinical Bacteriology, Karolinska Hospital, 104 01 Stockholm, Sweden*
NOVICK O. Mr.	*Israel Institute for Biological Research, Ness-Ziona, P.O. Box 19, Israel*
O'CONNEL D. C. Mr.	*Defence Research Establishment Suffield, Ralston, Alberta, Canada*
PARKER M. T. Dr.	*Central Public Health Laboratory, Colindale Avenue, London N.W.9*
PAVILANIS V. Dr.	*Virus Department, Institute of Microbiology and Hygiene, 531 Boulevard des Prairies, P.O. Box 100, Laval des Rapides, P.Q. Canada*
PEPYS J. Prof.	*Institute of Diseases of the Chest, Brompton Hospital, Brompton, London S.W.3*
PERKINS F. T. Dr.	*National Institute for Medical Research, Hampstead Laboratories, Holly Hill, London N.W.3*
POTTER, C. W. Dr.	*Department of Medical Microbiology, University of Sheffield, Sheffield*
PROFFETT S. Mr.	*Portex Ltd., Reachfield, Hythe, Kent*
RECHSTEINER J. Dr.	*Rijks Institut voor de Volksgezondheid, Sterrenbos 1, Utrecht, Netherlands*
ROBERTS R. H. Surg. Capt.	*Canadian Forces Medical Services, Office of Surgeon General, Canadian Forces, H.Q. Ottawa, Ontario, Canada*
RYLANDER R. Dr.	*Department of Environmental Hygiene, National Institute for Public Health, 10401, Stockholm 60, Sweden*

SALOMON L. Dr.	*Deseret Test Center, Building 103, Soldiers Circle, Fort Douglas, Utah 84113, U.S.A.*
SCHULMAN J. Dr.	*Cornell University Medical College, 130 York Avenue, New York, N.Y. 10021, U.S.A.*
SHANAHAN A. J. Dr.	*7217 Churchill Road, McLean, Virginia 22101, U.S.A.*
SHOOTER, R. A. Dr.	*Department of Bacteriology, St. Bartholomew's Hospital, London E.C.1*
SILVER, I. H. Mr.	*Microbiological Research Establishment, Porton, Salisbury, Wiltshire*
SMITH C. E. G. Dr.	*Microbiological Research Establishment, Porton, Salisbury, Wiltshire*
SMITH C. H. Dr.	*Evans Medical Ltd., Speke, Liverpool 24*
STREET B. K. Mr.	*Beecham Research Ltd., Clarendon Road, Worthing, Sussex*
STUART-HARRIS C. H. Prof.	*Department of Medicine, The Royal Hospital, Sheffield Sl 3SR*
SYKES G. Mr.	*Microbiology Division, Boots Pure Drug Co., West Bridgeford, Nottingham*
THOMAS G. Dr.	*Microbiological Research Establishment, Porton, Salisbury, Wiltshire*
THOMSON R. G. Prof.	*University of Guelph, Guelph, Ontario, Canada*
TOWERS A. G. Mrs.	*Cross-infection Reference Laboratory, Central Public Health Laboratory, Colindale Avenue, London N.W.9.*
TROWBORST T. Dr.	*Laboratory for Microbiology, Catharynesingel 59, Utrecht 6, Netherlands*
TYRRELL D. A. J. Dr.	*Common Cold Research Unit, Harvard Hospital, Coombe Road, Salisbury, Wiltshire*
VEDROS N. A. Dr.	*Naval Biological Laboratory, Naval Supply Center, Oakland, California 94625, U.S.A.*
WALDMAN R. H. Dr.	*Department of Medicine, Division of Infectious Diseases, University of Florida, Gainesville, U.S.A.*
WATKINS H. M. S. Dr.	*Naval Biological Laboratory, Naval Supply Center, Oakland, California, U.S.A.*
WENHAM R. B. Mr.	*Fisons Pharmaceuticals Ltd., Bakewell Road, Loughborough, Leics*
WESTWOOD J. C. N. Prof.	*Department of Microbiology and Immunology, Faculty of Medicine, University of Ottawa, 550 Cumberland Street, Ottawa 2, Ontario, Canada*
WILLIAMS R. E. O. Prof.	*Department of Bacteriology, St. Mary's Hospital Medical School, Paddington, London W.2*
WILLOUGHBY D. S. Dr.	*Bacteriology Section, Defence Research Establishment Suffield, Ralston, Alberta, Canada*
WINKLER K. C. Prof.	*Laboratorium voor Mikrobiologi de Rijksuniversiteit, Catharijnesingel 59, Utrecht, Netherlands*
WOLFE E. K. Mr.	*Aero and Evaluation Laboratory, Fort Detrick, Frederick, Maryland 21701, U.S.A.*

Foreword

The Third International Symposium on Aerobiology was the first to be held outside the United States. We were invited to hold it in the United Kingdom at the 1966 meeting in Chicago and this book demonstrates the success of the efforts of the Organizing Committee and its many willing helpers.

The objectives we set ourselves for the 1969 meeting were the promotion of discussion both inside and outside the meeting and it is my impression that both these objectives were achieved. The effort required to extract from authors the papers and abstracts to be presented at the sessions and seminars, sufficiently in advance of the symposium to permit pre-circulation, proved to be worth while and undoubtedly greatly contributed to the value of the discussion periods.

However, the main purpose of meetings such as this must be to bring into contact the aerobiologists and those who have a need for a knowledge of aerobiology. This, too, was achieved and many useful exchanges of information and opinion took place during the symposium and can confidently be expected to continue in the future.

The choice of the University of Sussex as the site for the symposium was particularly fortunate and the excellent facilities placed at our disposal were praised by everyone who took part. As we anticipated, the residential nature of the meeting greatly increased the opportunity for discussion. Our hope is that the organizers of the next symposium due to be held in 1972 will be equally fortunate in their choice. I am glad to be able to report that Professor K. C. Winkler of the Department of Microbiology, University of Utrecht, has been persuaded to accept responsibility for this next symposium which will, therefore, be held in the Netherlands. We wish him, and those who will be assisting him, every success in the Fourth International Symposium on Aerobiology.

C. E. GORDON SMITH
Chairman, Organizing Committee

Contents

CHAPTER I

SEMINAR **Immunological aspects of pathogenesis**

Chairman S. GARD *Convener* C. E. G. SMITH

CHAPTER II

SESSION **Value and use of germ-free environments**

Chairman R. BLOWERS *Convener* M. T. PARKER

SEMINAR **Vaccines and methods of vaccination**

Chairman F. T. PERKINS *Convener* F. T. PERKINS

Meningococcal transmission among army recruits

Malcolm S. Artenstein, M.D. and **Col. Thomas H. Lamson,** M.C.

Walter Reed Army Institute of Research, Washington, D.C., and Preventive Medicine Division, Fort Dix, New Jersey

Factors which affect the spread of meningococci from one person to another are poorly understood. Certain groups, however, seem to have a higher prevalence of nasopharyngeal meningococcal carriers than normal populations, especially families of meningitis cases and military recruits. Older studies have shown carrier rates in families of cases to vary from as low as 7 per cent to as great as 70 per cent (1). A recent study of a large family kinship in which several meningitis illnesses occurred showed 44 per cent of the 'tribe' to be carriers; 20 per cent of nonfamily contacts but only 3 per cent of other community members were carriers (2). Carrier rates in military populations may be variable, but among recruits in basic training rates of 85–90 per cent are frequently noted (3).

Close contact is a factor common to both of these populations. For many years crowding has been implicated as a major factor in meningococcal transmission; Glover (4), for example, observed a direct relation between bed distance and carrier rates (i.e. the closer the beds the higher the rate), although he also noted that exceptions may occur.

Seasonal variations in carrier rates as well as disease incidence have been documented (5) yet the precise factors (i.e. temperature, relative humidity) have not been adequately defined. In previous studies we have examined the relationship between acute respiratory disease or adenovirus infection and meningococcal carrier rates in Army recruits and found that no correlation existed (3). Furthermore, differences in types of barracks housing were not reflected in differences in meningococcal carrier rates.

The present investigations were undertaken to analyze the transmission of meningococci among Army recruits in an attempt to better define contact-transmission situations. In addition, a number of factors related to dissemination of organisms by carriers were studied.

Methods

Carrier status was detected using oropharyngeal swabbing as previously described (3). Briefly, use of selective medium containing lincomycin and polymixin allowed the isolation of meningococci in almost pure culture. Identification of strains was performed by classical fermentation and serogrouping procedures (6). The Litton Large Volume Air Sampler was used to measure dissemination of organisms by individual coughs (7).

Results

Transmission of meningococci among Army recruits sleeping in the same room. Company G-2 consisted of four platoons of 44–50 men each, of which platoons 1, 2 and 3 were housed in a three-storey brick building divided into two-bed and eight-bed rooms. Platoon 4 was housed in an old wooden open-bay structure in which 20 men slept in double-tiered bunks on the first floor and 28 on the second. During the fourth week of basic training two men on the second floor developed meningitis due to sulfadiazine-sensitive, group B meningococci. Nasopharyngeal survey performed one week later showed that the percentages of men who carried meningococci were similar in all platoons and that no differences were apparent between men in the fourth platoon who slept on the first floor and those who slept on the second (Table 1). Furthermore, in this platoon, distance of a man's bed from the index cases bore no relationship to carrier state.

During the next three weeks the percentage of men who were carriers doubled in the company and at this time 28 per cent of men in the fourth platoon were carrying an organism similar to that which caused the two illnesses. Only 10 per cent of men in the other three platoons carried a similar meningococcus. Thus, in this circumstance, the virulent strain appeared to disseminate widely in the fourth platoon but after removal of the two index cases.

Analysis was made to determine how often recruits acquired meningococci from their room mates. At the time of the first survey, there were five two-man rooms in which one of the two occupants was a carrier. Men from four of these rooms were included in the survey taken three weeks later. In three instances there was no acquisition by the other man; the other occupant of the fourth room acquired a meningococcus different from that of his room mate. In 15 rooms occupied by seven or eight men each, acquisitions of meningococci unlike index strains were more frequent than those of index strains (Table 2).

A similar study conducted in another company one year later (Table 3) showed similar results. Thus, less than half the acquisitions were derived from room mates. Although the strain markers used ((1) grouping by means of antiserum and (2) sulfadiazine sensitivity or resistance) lack the precision necessary for definitive epidemiologic study of person to person spread, observation of the social habits of recruits has shown that many opportunities for close contact occur with other recruits within the platoon and other platoons which could serve to transmit meningococci. These results indicate that attempts to prevent transmission by manipulating sleeping quarters cannot be expected to be totally effective.

Dissemination of meningococci by carriers

Aerosols produced by coughing were collected in a standard fashion by methods previously described (7). Single coughs were utilized after preliminary studies showed that the first of two coughs usually produced the larger number of recoveries. Comparison of cough samplings from individuals tested at one- or two-day intervals revealed minimal variations in dispersal. A number of factors which might affect dispersal of meningococci were evaluated. (1) Comparison of respiratory tract symptoms and meningococcal dispersal is shown in Table 4. In each instance symptoms were mild and only one throat culture yielded a viral agent (herpes simplex). No correlation was found between presence of ARD and magnitude of dissemination. In a previous paper we have reported that simultaneous infection with an adenovirus was not associated with increased meningococcal dispersal (7). (2) Group C, sulfadiazine-resistant strains have been the cause of most of the meningococcal disease in Army personnel in the past year whereas nongroupable strains have not been isolated from patients with systemic illness. Therefore, serogroup and/or sulfadiazine sensitivity of meningococci were compared with magnitude of dispersal. From Tables 5 and 6 it is apparent that group C and sulfa-resistant characteristics did not correlate with increased dispersal. In this company group C strains and nongroupable meningococci spread to about equal degree (31 men acquired a group C and 35 acquired a nongroupable strain); however, during the subsequent five months group C sulfadiazine-resistant strains accounted for the great majority of strains isolated from carriers at Fort Dix. Relative humidity (RH) and temperature were fairly constant during the period of cough sampling: RH varied from 30 per cent to 51 per cent; temperature ranged from 68° to 75°F. Thus, the effect of these two environmental factors upon dissemination could not be evaluated.

The relationship of duration of the carrier state and magnitude of cough dispersal was assessed by comparing cough data from two groups of carriers. Meningococcal pharyngeal acquisitions among Army recruits were determined by weekly carrier surveys. Laboratory personnel were cultured intermittently over a period of months. Cough samples were obtained on three or more occasions, but the data in Table 7 are based upon the cough which yielded the maximum number of organisms for each carrier. From Table 7 it can be seen that, in general, chronic carriers (duration greater than eight weeks) are poor dispersers. Only 3 of 12 individuals expelled more than 25 organisms per cough and, it should be emphasized, this number represents but one cough per person at this level of excretion; many other samples produced less than 25 organisms. Newly acquired carriers were more likely (8 of 16 persons) to disperse over 25 organisms per cough and, generally, this level of dispersal remained constant over several days of testing. Further data are needed to establish this point conclusively.

A summary of the various factors related to dispersal of meningococci by coughing is shown in Table 8.

Local immunity to the meningococcal carrier state

The development of purified meningococcal polysaccharide vaccines has been recently reported (8). In addition to stimulating the formation of serum antibodies, the ability of the group C vaccine to interrupt transmission was demonstrated in a study in three companies of Army recruits in which approximately 20 per cent of the men were vaccinated (9). Table 9 shows the reduced carrier rates for group C meningococci among vaccinees compared to controls. Thus, this vaccine appears to produce a local nasopharyngeal immunity to colonization by the meningococcus. It appears likely that immunization of a larger proportion of a population would reduce transmission even more strikingly and in this fashion reduce the exposure of susceptible recruits.

Discussion

Most meningococcal illnesses are acquired from asymptomatic carriers rather than from overtly sick individuals (10). Because of the low attack rates for disease, however, carrier studies provide a better index of transmission of the organism. The present studies have shown that although many meningococcal acquisitions among recruits can be related to sleeping quarters, at least as many were not derived from room mates. Therefore, attempts at environmental control of transmission will be unsuccessful if only barracks contacts are reduced. Attempts to reduce meningococcal transmission indirectly by methods aimed at reducing respiratory infections, which are often associated in time with meningococcal infections, are not likely to be effective since the magnitude of dissemination is unrelated to respiratory tract symptoms. Although it could be argued that recruits with ARD cough and sneeze more each day than normals and thus would contaminate more contacts, previous studies have shown no relationship of ARD and meningococcal carrier rates (3).

From the data presented here it would appear that the development of potent vaccines offers a more satisfactory means for prevention of meningococcal disease than environmental methods. Vaccines currently under study produce humoral antibody, local respiratory tract immunity and a reduction in transmission among immunized populations.

Summary

Meningococcal carrier surveys among Army recruits demonstrated that sleeping quarters are a source of acquired strains, but not the only important source. The magnitude of dispersal of meningococci by single coughs is not increased by simultaneous upper respiratory illness. Microbiological characteristics such as specific serogroup and sulfadiazine resistance do not correlate with dispersal. Chronic carriers are poor dispersers compared to newly acquired carriers. Polysaccharide vaccines appear to offer a more satisfactory means of controlling transmission than environmental manipulations.

Acknowledgements

Assistance with the air-sampling studies was provided by Dr. William Hankins and the staff of the Experimental Aerobiology Division, U.S. Army Biological Laboratory, Ft. Detrick, Maryland. K. O'Brien, J. Funk, C. Harkins and B. Brandt provided valuable technical assistance.

REFERENCES

1. Worster-Drought, C. and Kennedy, A. M. (1919). *Cerebrospinal Fever*, A. & C. Black Ltd., London.
2. Greenfield, S. and Feldman, H. A. (1967). Familial carriers and meningococcal meningitis, *New England J. Med.*, **277,** 497–502.
3. Artenstein, M. S., Rust, J. H., Hunter, D. H., Lamson, T. H. and Buescher, E. L. (1967). Acute respiratory disease and meningococcal infection in Army recruits, *J. Amer. Med. Assoc.*, **201,** 1004–1008.
4. Glover, J. A. (1920). *Cerebrospinal Fever: Studies in the bacteriology, preventive control and specific treatment of cerebrospinal fever among the military forces, 1915–1919*, Med. Res. Council, Spec. Rep. Series No. 50, London, His Majesty's Stationery Office, Section 20, pp. 133.
5. Aycock, W. L. and Mueller, J. H. (1950). Meningococcus carrier rates and meningitis incidence, *Bact. Rev.*, **14,** 115–160.
6. Evans, J. R., Artenstein, M. S. and Hunter, D. H. (1968). Prevalence of meningococcal serogroups and description of three new groups, *Amer. J. Epid.*, **87,** 643–646.
7. Artenstein, M. S., Miller, W. S., Lamson, T. H. and Brandt, B. L. (1968). Large-volume air sampling for meningococci and adenoviruses, *Amer. J. Epid.*, **87,** 567–577.
8. Gotschlich, E. C., Goldschneider, I. and Artenstein, M. S. (1969). Human immunity to the meningococcus. IV. Immunogenicity of group A and C meningococcal polysaccharides in human volunteers, *J. Exp. Med.*, **129,** 1367–1384.
9. Gotschlich, E. C., Goldschneider, I. and Artenstein, M. S. (1969). Human immunity to the meningococcus. V. The effect of immunization with meningococcal group C polysaccharide on the carrier state, *J. Exp. Med.*, **129,** 1385–1395.
10. Murray, E. G. D. (1929). *The Meningococcus*, Med. Res. Council, Spec. Rep. Series No. 124, London, His Majesty's Stationery Office.

TABLE 1

Meningococcal carrier surveys, Co. G-2

		Date of survey			
Platoon		17 Nov.		6 Dec.	
1, 2, 3		34/118[a] 29%	8B[b]	88/141 62%	14B
4	First floor	5/19	2B	10/19	5B
	Second floor	4/22	1B	16/27	8B
		9/41 22%	3B	26/43 57%	13B

[a] No. men positive/no. men tested.
[b] Group B, sulfadiazine-sensitive strains.

TABLE 2

Secondary attack rates for carrier state in 7–8 man rooms in a three-week period (Co. G-2)

29 index carriers:	2 converted to another index strain
	6 acquired a new strain
51 noncarriers:	13 (26%) acquired an index strain
	15 (29%) acquired a new strain
	2 (4%) acquired a nongroupable strain

TABLE 3

Secondary attack rates for carrier state in 7-man rooms in a three-week period (Co. A-1–3)

41 index carriers	
77 noncarriers:	6 (8%) acquired an index strain
	19 (25%) acquired a new strain
	13 (17%) acquired a nongroupable strain

TABLE 4

Relation of number of organisms dispersed to presence of acute respiratory disease (ARD)

	No. of individuals with	
No. of organisms	*No ARD*	*ARD*
0–12	7	5
13–25	1	0
26–50	1	3
51–100	1	3
101–200	1	1

TABLE 5

Comparison of meningococcal serogroup carried and number of organisms dispersed

	No. of organisms				
Serogroup	*0–12*	*13–25*	*26–50*	*51–100*	*101–200*
NT	4[a]	0	0	1	1
B	3	0	2	0	0
C	5	1	2	2	1
Bo	2	0	0	0	0
29E	0	0	0	1	0

[a] No. of men, based on initial cough sample.

TABLE 6

Relation of sulfa resistance to dispersal

	No. of organisms				
Sulfonamide	*0–12*	*13–25*	*26–50*	*51–100*	*101–200*
Resistant	6[a]	0	3	2	1
Sensitive	8	1	1	2	1

[a] No. of men, based on initial cough sample.

TABLE 7

Duration of carrier state compared with magnitude of dispersal

	No. of organisms				
Duration	*0–12*	*13–25*	*26–50*	*51–100*	*101–200*
Less than 2 weeks	4[a]	0	1	2	3
2–3 weeks	3	0	0	0	0
Less than 8 weeks	1	0	0	1	1
More than 8 weeks (recruits)	1	0	0	1	0
More than 8 weeks (Lab. personnel)	5	3	1	0	1

[a] No. of men based upon maximum cough.

TABLE 8

Factors related to magnitude of dispersal

Factor	*Conclusion*
1. Magnitude of nasopharyngeal swab culture	Not related
2. Simultaneous viral infection	Not related
3. Presence of respiratory symptoms	Not related
4. Serogroup of organism	Not related
5. Sulfadiazine resistance	Not related
6. Variation in dispersal from day to day	Minimal
7. Duration of carrier state	Suggestive
8. Ambient temperature	Not tested
9. Ambient relative humidity	Not tested
10. Viscosity of secretions	Not tested
11. Presence of local antibodies	Not tested

TABLE 9

Percentage of recruits who became carriers of group C meningococcus

Company	*Controls*	*A vaccine*	*C vaccine*
B-6-3	42	37	24[a]
E-5-3	38	—	4·6[b]
E-2-3	69	68	31[b]

[a] P = < ·05. [b] P = < ·0001.
(From Gotschlich *et al.*, Ref. 9.)

Epidemiologic Investigations in Polaris Submarines

H. M. S. Watkins *et al.*

Naval Biological Laboratory, Oakland, California

The advent of nuclear submarines and of space vehicles has introduced mankind to an environment with which humankind has had no prior experience. In these vehicles man has taken the first steps towards sealing himself for indefinitely prolonged periods in closed, recycled, artificially replenished atmospheres. In all his previous history man has been exposed to such environments for a few hours or days at most; now he is subjected to a biological impact of unknown consequence on a round-the-clock basis for many weeks; tomorrow, as space and deep submergence exploration carry man even further from his customary habitat, he may live under such conditions for years.

During the development of nuclear submarines and of space vehicles extensive investigations have been made to redefine physiologic tolerances (1) and to identify toxic hazards (2, 3) imposed by confinement in closed cabin environments. However, very little work has been devoted to the microbial ecology, to the possible alterations in infectious processes, affecting men in confined atmospheres. Particularly we should learn whether the changes observed in host-parasite relationships of populations isolated on islands (4, 5) or in the Antarctic (6–8) will be duplicated on crews of these vehicles.

With such questions in mind, intermittently during the years 1962 to 1967 the staff of the Naval Biological Laboratory (NBL) of the University of California, in collaboration with submarine medical officers, has studied airborne microbial contamination and respiratory infections occurring in Polaris submarine crews at sea. This report summarizes the principal findings.

Polaris environment and operational cycle

In a Polaris submarine on patrol, a population of 130 to 140 highly motivated, physically select, healthy young men are confined in a closed, congested space for approximately two months during which the submarine ordinarily never surfaces. The air environment is completely recycled and regenerated. Devices to remove atmospheric chemical and organic contaminants include inert filters to remove dust particles, activated charcoal filters, electrostatic precipitators, carbon monoxide burners, and carbon dioxide scrubbers. All of these to some extent also removed airborne microbial contaminants, but their overall efficiency in this respect had not been extensively examined prior to these studies.

Each Polaris submarine has two alternating complete crews, the Blue crew and the Gold crew. Following shake-down and training cruises, a new Polaris submarine makes its first extended patrol manned by the Blue crew, enroute to its overseas base. Each base has a submarine tender as a supply ship and a complete floating repair facility. Submarine crews in harbor have ready access to local recreational areas and some members of the supply and repair crews may reside ashore.

Three or four days prior to the arrival of the submarine at the overseas base, the Gold crew is flown directly from the United States and is housed aboard the tender. Upon arrival of the submarine, the Blue and Gold crews intermingle for several days, following which the Blue crew is flown to the United States where crew members disperse to their homes on leave. The Gold crew is now moved aboard the submarine; daily contact continues between crew members of the submarine and the tender until the submarine departs on patrol.

Accordingly, a Polaris submarine crew immediately before diving has ample opportunity to become seeded with the microbial flora present in its home port in the United States, in the local inhabitants at an overseas base, or in the alternate crew returning from patrol.

Outline of studies

These investigations were conducted in four phases, Phases I and II were conducted in the Atlantic, Phase III in the Pacific Ocean.

Phase I. In 1961, the Naval Biological Laboratory was asked to study the content and spread of micro-organisms in the air and crew of a Polaris submarine (Sub 'A') during prolonged submergence. Since severe spatial restrictions precluded isolation and identification of micro-organisms at sea, specimens were stored in two mechanical deep freezers installed in a submarine while it was being constructed. After completion of a patrol nominally lasting 60 days, specimens were sealed in cans for air shipment to NBL at Oakland, California, where they were subjected to microbial examination. NBL staff (CEM or JPH) participated in three successive patrols with the Blue crew of this test submarine in 1962.

Phase II. More extensive studies were made during 1964–1965 in another submarine of newer design (Sub 'B'). This vehicle was accompanied by NBL staff (MAM, JPH or TRW) on five successive patrols. Studies in each of these submarines were terminated when deep freezers were removed. This is worthy of note since the loss of our capacity to freeze large numbers of samples resulting from the extrusion of our dismantled deep freezer through a 26-inch diameter submarine hatch effectively, although regrettably, settled any question of repeated or confirmatory experimentation of this type in the same submarine at a later date.

Phase III. In 1966–1967, data were collected from several submarines operating out of Guam. By this date we had developed technics and miniaturized portable deep freezers and roller drum incubators (Figures 1, 2, 3) which permitted some microbial isolation procedures on a limited scale to be conducted at sea. In this portion of the studies, submarine medical officers spent two-week intensive indoctrination periods at NBL prior to a sea patrol on which they independently conducted specified studies and collected data and specimens which subsequently were further processed and evaluated at NBL.

Phase IV. Finally, statistical data were compiled from over 100 reports of medical officers on Polaris submarine patrols to permit comparison of some of our findings on respiratory infections with reported experience in the Polaris submarine force at large.

Procedures and equipment

1. *Protocol booklet.* Each investigator was provided with a booklet which listed supplies and equipment provided; examples of each record form required; detailed protocols and flow charts of each procedure to be carried out at sea; coding, packaging and shipping instructions; and a summary, describing the study, to be used as an information sheet for the crew. Contents of booklets varied in accordance with the differing objectives of each patrol.

2. *General procedure.* NBL investigators joined submarine crews approximately one or two weeks before deployment to sea from the overseas base. Medical officer investigators, by contrast, accompanied crews during flight to the overseas port and accordingly were present during the one-month refit period prior to deployment to sea. Completion of questionnaires during flight and adherence to experimental protocols at sea were usually successful as planned; however, great difficulty occurred in collecting data and repetitious specimens during the refit periods when all submarine and tender crew members work under severe pressure to ready the ship for sea.

On some patrols clinical specimens were collected from the entire crew; on others, particularly when air samples were simultaneously collected, test groups of 30 or 50 men were studied. Serum specimens were collected one or two days before or after deployment to sea, and again during the last week at sea; where appropriate, serum was also collected from the few crew members who reported to Sick Bay with respiratory infections. Oropharyngeal gargle, and skin, nasal or throat swabs were usually collected semi-weekly at the beginning of a patrol and weekly after the second or third week at sea. At this time upper respiratory infection (URI) symptoms were coded and recorded by the investigator as nasal congestion, sore throat or hoarseness, cough (nonproductive), cough (productive), or systemic such as fever, aches, malaise, etc.; in addition, an URI check-off sheet was posted on which the entire crew recorded presence or absence of URI symptoms once or twice weekly. During the Phase II studies (Sub 'B'), medical histories were abstracted from each crew member's records; a special form was completed on each crew member reporting to Sick Bay for whatever cause; a detailed epidemiologic questionnaire was administered; and bunk location charts were prepared for all crew members. No attempt was made to group crew members by work assignment or location because, unlike other Naval ships, it was determined that intermingling during messing, and in movies, washrooms and bunking areas, would make such groups meaningless.

3. *Gargle and swab specimens.* During Phase I and II studies, 5 per cent milk in saline was employed as a protective medium for bacterial and viral specimens collected by oropharyngeal gargle, by a moistened swab applied to skin or nares, or by a dry swab applied to the oropharynx. In later studies, dry swabs were immersed and broken off in veal-infusion broth or in L-15 tissue-culture medium in some cases in parallel with skim-milk specimens. Swabs and collection fluids in one-dram, screw-cap glass vials were promptly stored at reduced temperature in deep freezers for duration of the patrol.

During Phases I and II, built-in deep freezers operated at −70°C or lower temperatures; portable miniaturized deep freezers employed in Phase III operated at −60°C. However, it should be noted that at certain times freezers were subjected to temporary loss of power, with resultant cycling of temperatures of samples and specimens in the frozen state into the −20° to −30°C range known to be detrimental to microbial viability. Confirmation that thawing did not occur was established by the incorporation of recording thermometers in freezers and frozen specimen shipments.

On reaching port, frozen specimens were hermetically sealed in tin cans with the aid of a commercial canning device, and were air shipped on dry ice to NBL. Specimens collected from symptomatic individuals during the latter part of Phase II, and during Phase III, customarily were processed for isolation of viruses within two weeks after arrival; however, limitations of laboratory staff and facilities required maintenance of many other specimens at −70°C for as much as three months after arrival.

4. *Air samples.* Air samples were collected at a specified time daily or on alternate days throughout patrols in Phase I and during four patrols of Phase II at eight or more sampling areas representative of major crew activities (i.e. mess area, berthing area, washroom, etc.). In addition, at several stations air samples were collected at six-hour intervals, or before and after crew operations which were expected to produce an increase in airborne microbial contamination. Figure 4 shows a typical impinger-pump installation located above a wash basin in the crew's lavatory.

Air samples were collected with the aid of metallic, raised-jet air impingers equipped with a 6·6 litres/min. restrictive orifice. Samples were collected for 20 min. in 5 per cent skim-milk saline containing 0·1 per cent antifoam compound; larger sampling times caused excessive fluid evaporation. Impingers were operated by high-speed vacuum pumps equipped with automatic timing devices. Samples were divided into three aliquots which were stored frozen as noted above.

5. *Filters on sanitary and waste tank vents.* During three patrols in Phase II, high-efficiency particulate filters with bypass were installed and tested for effect on levels of air-microbial contaminants during the first half of each patrol.

6. *Bacterial procedures.* Air samples and skin and nasal swabs were examined for viable hemolytic and enteric bacteria. Although some variations occurred in the course of the studies, the following media and procedures were customarily employed: viable counts and hemolytic counts were made by surface plating sample aliquots respectively on heart infusion agar and on blood agar base containing 5 per cent sheep cells. Enterobacteriaceae were determined by biochemical and fermentation reactions of colonies picked from eosin methylene blue agar plates. Fecal enterococci were isolated by growth in enterococcal presumptive broth streaked on mitis-salivarius agar, with subsequent characterization of colonies by Sherman's criteria (9). Coagulase-positive staphylococci, isolated on the media described by Finegold and Sweeny (10), were typed in accordance with methods described by Blair and Williams (11), utilizing typing phages prepared by the Communicable Disease Centre, Atlanta, Georgia.

7. *Viral procedures.* Isolation of viruses from specimens and air samples was primarily carried out at NBL. However, for Phase II patrols, isolations were also made from specimen aliquots sent to the Sixth Army Medical Laboratory (6th AML), Fort Baker, California, and during one patrol of Phase II and two patrols of Phase III, immediate inoculations and incubation of a few specimens in tissue cultures were carried out at sea or on a tender. Identification of isolates was initiated at NBL and completed at 6th AML.

Each specimen or sample was passed, up to three times, in a minimum of three cell systems: cells employed included established embryonic rhesus monkey kidney (EMK), embryonic human kidney (HEK), embryonic human lung (HEL), embryonic human skin and muscle (HESM), and adult rhesus primary monkey kidney (MK), Hep-2 and WI-38. Diploid cell systems were developed at NBL (HEK) or were obtained from commercial sources. All cells were propagated in Eagle's Minimal Essential Medium (MEM) or in Medium L-15, plus glutamine, 10 per cent inactivated foetal calf serum (FCS) and antibiotics. The maintenance medium was L-15 plus 2 per cent FCS. Inoculated cells were incubated in roller drums at 34°C.

Cells to be used at sea or on an overseas submarine tender were usually transported at ambient temperatures by air from NBL as preformed monolayers in one-dram screw-capped glass vials. In several Phase III patrols, frozen seed cell suspensions were air shipped in sealed cans on dry ice and subsequently employed to produce monolayers, utilizing L-15 medium for growth and maintenance in vials, and a portable, miniaturized, roller drum incubator and deep freezer for subsequent incubation and storage of cultures. Additional details of these field isolation methods will be published elsewhere (12).

Cultures were routinely tested for hemadsorption with human type 0 erythrocytes after 7 to 10 days' incubation, and were passed after 14 days. Rhinovirus isolates were identified on the basis of thermal, acid, chloroform, and 5-iododeoxyuridine stability; systematic typing was usually omitted. Other isolates were identified at 6th AML by neutralization with commercial antisera.

8. *Serological determinations.* Cross-identity of Rhinovirus isolates was established by neutralization in WI-38 cells after virus adaptation, using inactivated guinea-pig antisera produced at NBL. Neutralization of isolates by inactivated paired sera from crew members was carried out, usually with 10^2 $TCID_{50}$ of virus and serially diluted sera, after incubation at 37°C for two hours.

Sera from several patrols were screened for micro-complement fixation titres, using two antigen and two complement units with overnight fixation at 4°C. Commercial antigens were used for Influenza A, B, C; Parainfluenza 1, 2, 3; Adenovirus; Herpes simplex; Mumps; Respiratory Syncytial; and in some cases Poliovirus 1, 2, 3.

Results

1. *Collection medium.* Since respiratory and other airborne bacteria recovered in these studies had to withstand frozen storage for periods in excess of 70 days, a preliminary survey was made of 32 preparations thought suitable for maintaining viability of oral bacteria in oropharyngeal saline gargle specimens, or of pure cultures of *Staphylococcus aureus,* during prolonged storage at 4°C, −20°C, and −70°C.

Table 1 lists the preparations tested, and notes six preparations which permitted nominal 100 per cent viable recovery after six weeks at −70°C, when stored preparations were plated out on either blood agar base or on 1·5 per cent sheep blood agar. At 4°C microbial growth occurred, and at −20°C counts were highly variable Therefore, specimen storage required use of deep freezers. Skim milk, 5 per cent in saline, was selected for use, in part because of its simplicity and relative palatability. A brief trial indicated that this medium also could be used satisfactorily, with added 0·1 per cent Dow Corning Antifoam, in impingers for collection of airborne bacteria.

2. *Airborne bacteria on patrol.* Air samples collected from all three patrols in the first submarine studied usually yielded low viable counts at all sampling stations and at all four times in the day when samples were routinely collected. Figure 5 represents a schematic presentation of the major living and working spaces in this class of nuclear submarine, and lists the areas where samples were collected, namely the torpedo compartment, the mess area and galley, the crew's lavatories and berthing areas, and the missile compartments.

Figure 6 shows a cross-sectional view of the middle and lower levels in the submarine and again shows the location of six of the eight air sampling stations. It also illustrates the geographical distribution of the sleeping areas: with officers located one, two or three to a stateroom; with semi-isolated areas on the middle and lower level for the two groups of chief petty officers; and with tiers of bunks three high for occupancy by the crew. This normally provides sleeping space for a total of 140 men.

Mean airborne counts during two patrols from seven sampling stations indicated that, overall, the bacterial burden of air was maintained at approximately 10 micro-organisms or less per litre (Figure 7). Collection of samples at different times of the day or at different sampling stations made little difference, although there was a small but real increase ($p < 0{\cdot}01$) in bacterial content of the air in areas where there was greater activity of crew members. No significant buildup in airborne bacteria was detected as a patrol progressed.

However, on numerous occasions during each of the three patrols, air counts in excess of 30,000 organisms/$ft.^3$ were detected. This occurred 51 times during the second patrol, as is shown in Table 2; furthermore, 24 of these 51 events occurred in a total of only five days. On each such day high counts were obtained simultaneously at four or more stations in the submarine, suggesting that the source of these high concentrations was not a counting artifact or a localized sampling event. Since the sampling method at a given station collected air only during 20 min. in each six hours, we do not know how long a high level of contamination persisted; however, it may be noted that only once were high levels obtained at two successive sampling times, from which we concluded that the normal air purification devices in the submarine were adequate to remove the air contaminants in a matter of several hours, if not sooner.

Evidence that air contaminants were potential human pathogens was sought by examining air samples, skin (nape of neck) and nasal swab specimens, and oro-pharyngeal gargle specimens for presence of coagulase-positive staphylococci, enterococci and other enteric bacteria. Table 3 summarizes the numbers of samples positive for these micro-organisms. They were recovered in low numbers in all types of specimens collected throughout a patrol, suggesting the presence of continuing or intermittent exposure to low-grade airborne contamination with staphylococci and fecal organisms. No enteric pathogens were isolated.

The obvious suspect source of fecal contaminants was the sanitary waste system of the submarine. Sanitary wastes and wash waters are collected in tanks which are periodically evacuated overside by means of air under high pressure. Since this air cannot be allowed to escape outboard, it must be discharged inboard before the emptied tank can be further employed. This is done by bleeding the air through a critical orifice and filter system into the living spaces of the submarine. However, in the third patrol attempts to associate high levels of airborne contamination with the inboard venting times for the sanitary waste tanks were unsuccessful; similarly, no unusually high counts were obtained in samples collected during the venting periods.

Nevertheless, it was considered justified to attempt to control possible bacterial discharges from inboard sanitary vents by installation of filter canisters containing asbestos fibre packs rated for 99·97 per cent collection of submicron DOP particles with a 3 $ft.^3$/min. air flow rate. These filters were employed on all vents during the first three to four weeks of two patrols (V and VI) in the second submarine studied, while air samples were collected at 3 to 16 sampling stations as before. It was hoped that the filters would contain the high 'breakthrough' counts obtained in the first submarine; however, the experiment was a failure because in five successive patrols of the second submarine of more modern design, with or without use of filters, no high counts were recorded.

Table 4 summarizes the air microbial counts obtained at two sampling stations, namely one in the crew's lavatory area and one in a main passageway, during one patrol of Sub 'A' and four of Sub 'B'. The larger percentage of high counts in the first submarine is evident, but more important is the observation that at about 80 per cent of sampling times in the first submarine and at about 95 per cent in the second, levels of microbial air contamination were maintained at 10 micro-organisms or less per litre of air. Since Patrol IV, in actuality, was the first patrol of a newly built submarine, the data suggest a trend towards improved air sanitation in its successive patrols. Inspection of the detailed data for each patrol in this submarine again indicated no buildup in air contamination as a patrol progressed.

3. *Transmission of staphylococci in submarine environment.* We wished to determine whether the closed submarine environment induced any detectable changes in the transmission of micro-organisms among crew members, or in the continued colonization of a bacterial carrier. Although no staphylococcal disease occurred during these patrols, we sought to evaluate transmission or disappearance of coagulase-positive staphylococcal strains by phage-typing organisms recovered on nasal or skin swabs. In a preliminary trial, 30 men on Patrol II were swabbed biweekly 17 times; in a later trial, the entire crew of 145 men on Patrol IV were swabbed weekly eight times. The findings are summarized in Tables 5 and 6.

The results of the first trial (Patrol II, Table 5) indicated that the typing procedure could successfully be applied to cultures of frozen specimens shipped to NBL; they also suggested that approximately one-third of the men were free of coagulase-positive staphylococci although eight men were persistent carriers of strains belonging to four phage types. Five recoveries were made that suggested transient acquisition of a different type strain; however, only three men appeared to acquire and retain a strain after yielding negative cultures for several weeks. Conversely, after two weeks into the patrol, only one man lost a resident strain and became culturally negative.

Essentially similar data were obtained from cultures collected from the entire crew of Patrol IV (Table 6): half the crew were persistently negative, and one-quarter were nasal carriers of strains predominantly grouped in five phage-type patterns. Only three men became negative after three consecutive weeks of harbouring a strain, and only one became persistently positive in mid-patrol. However, in five instances, nonresident phage-types were isolated from persistent carriers of a different phage-type. Replicate isolations were not attempted, so we have not determined whether these five men temporarily had a double infection, as would seem likely. There was no substantial change in the numbers of crew members culturally positive in succeeding weeks of the patrol: positive cultures were obtained from 45, 46, 47, 44, 39, 43 and 39 men in successive weeks.

4. *Respiratory infections: prevalence and characteristics.* No severe respiratory disease was seen during the patrols of Sub 'A'; however, an apparent common cold syndrome was widespread during the first weeks of each patrol. These observations for Patrols II and III are illustrated in Figure 8, the data for which were elicited by

individual biweekly questioning of test groups of 30 men, augmented by a posted semi-weekly symptomatology check-off list completed by the entire crew. Symptoms recorded included nasal congestion, sore throat or hoarseness, productive or nonproductive cough, and systematic complaints such as aches and pains and elevated temperature.

It is apparent from Figure 8 that the data elicited from the test group were representative of those acknowledged by the crew as a whole. In Patrol II, approximately 25 per cent of the crew reported common cold symptoms at the start of the patrol; this increased to two-thirds of the crew after approximately 10 days, and then gradually decreased to about 30 per cent after three to four weeks and only 10 per cent after six weeks.

Patrol III afforded a higher initial prevalence of cold symptoms with one-third of the crew affected at the beginning of the voyage, and 50 per cent affected after seven days at sea. By the fourth week only 10 per cent were reporting symptoms. At this time, the submarine surfaced in harbour for two days, and the crew had contact, to a limited extent, with the crew of a submarine tender before proceeding again to sea. One week later, that is at the end of the fifth week of this patrol, a second flare-up had occurred, with approximately 30 per cent of the crew reporting symptoms. By the sixth week, this had again subsided, with only 10 per cent of the crew still being affected.

A scatter diagram (Figure 9) represents the duration of symptoms as reported by the test group of 30 men during the second patrol. Eight men were asymptomatic at the beginning of the patrol and remained so throughout the resultant voyage of 46 days. Seven men came aboard with cold symptoms, and continued to be afflicted with them for periods ranging from 10 to 46 days. The median duration of symptoms in this group was three weeks. An additional 10 men became affected by the sixth day of the patrol, the median duration of symptoms in this secondary group being nine days. The third group of four men reported themselves symptomatic on the tenth day with a median duration of four days. After the first two weeks of the patrol, in only two instances was there any subjective evidence of further infection; in both cases these apparent new colds were of short duration and occurred on the 35th and 38th day of the patrol, respectively.

It appears, then, that there was a marked increase in mild respiratory distress, affecting from one-half to two-thirds of the crews of this ship during early stages of a patrol. It may be further noted that Patrol II and III occurred, respectively, in September to November and in March to May, sailing from Holy Loch, Scotland.

Approximately two years after Patrol III, we started a second series in Sub 'B', sailing out of Rota, Spain. During this time the refit period in harbour had been extended from two to three weeks to a full four weeks; since that time the peak of respiratory infections usually has occurred during the refit, prior to sailing.

At that time the significance of the refit—a time of great strain for a crew introduced precipitately to a different environment overseas, often immediately after a last night 'on the town', living at first on the tender under objectionably crowded and climatically uncomfortable conditions, stressed by a tremendous volume of work preparing the boat for sea—was not fully appreciated, and no detailed records of prevalence of mild respiratory infections were prepared. It must be appreciated that men under these conditions commonly do not report to Sick Bay for treatment, so these mild or chronic infections as a rule appear in the medical records of a patrol only in terms of comments by the medical officer. However, conversations with such medical officers clearly indicate that this refit period, in which a crew is seeded with the microbial flora of the tender crew and local population ashore at the overseas harbor, in addition to that introduced from home communities in the United States, governs the pattern of upper respiratory infections customarily seen on patrol.

Usually these take the form illustrated in Figure 10, which represents the prevalence, onset and duration of mild respiratory symptoms recorded on a patrol that left for sea from Rota during the month of March. At the patrol start, approximately one-half the crew had an upper respiratory infection; this number decreased progressively so that only five men reported symptoms in the eighth week of the patrol. However, the data suggest that new infections and reinfections, albeit in steadily decreasing numbers each week, did occur throughout the patrol.

Similar patterns of mild upper respiratory infections, together with recrudescence after two-day visits to harbor, were observed in three other patrols of Sub 'B' sailing at different months out of Rota, and have also been reported by our collaborating medical officers on submarines sailing from Guam in the Pacific Ocean.

However, another pattern of respiratory infections was seen in Patrol IV. A moderately severe, afebrile respiratory syndrome appeared during the three-week period following the 20th day at sea. This was marked by increased frequency and severity of sore throats and fatigue, nasal congestion, and some systemic pains and headache. Ten to 15 men each week reported to Sick Bay for treatment, a point of differentiation from URI seen earlier. The illness affected 21 men who had not had URI symptoms earlier in the patrol and 13 who had. It had a mean duration of one week, with no observed sequellae. It affords the only example of so-called 'mid-patrol syndrome' to occur on submarines we have studied, and it serves to illustrate the degree to which Sick Bay reports of illness may mask the total amount of respiratory symptomatology present in a crew. In Figure 11 we have plotted, at approximately weekly intervals, the total URI symptoms elicited by questionnaire, total men reporting to Sick Bay for any cause and for URI complaints, and elicited complaints of systemic distress. The last is a symptom uncommonly recorded in the common cold syndrome previously described. From these data it appears that the Sick Bay reports recorded less than half the number of crew members affected by this URI outbreak.

5. *Etiological studies: serological surveys.* During the patrols we accompanied in Sub 'B', paired sera were collected either from the entire crew, or from a test group of 30 men. The first serum was usually collected within 48 hours of the start of a patrol, the second over several days during the last week at sea. In later patrols in other submarines (Phase III), medical officers collected acute sera individually at the time

specimens were taken for viral isolation. Accordingly, the paired sera employed in the surveys were collected nominally seven weeks apart; whereas with paired sera collected from individuals reporting with URI, the acute serum may have been collected anywhere from seven to ten weeks prior to the convalescent serum.

The initial serum survey was made by micro-complement fixation on 127 paired sera collected from Patrol IV, in an attempt to establish the etiology of the mid-patrol outbreak. Sera were tested with 13 viral antigens and with *Mycoplasma pneumoniae.* Titres obtained are listed in Table 7. No diagnostic titre increases ($\geqslant$4X) were noted with any of the antigens employed. Subsequently, titres in the sera of 49 of these men were compared with those of sera collected approximately six months prior to and six months after Patrol IV. This comparison yielded diagnostic titre increases in two individuals for Influenza C virus and in five men for *M. pneumoniae*; all such increases occurred in the months prior to Patrol IV.

A similar survey was made subsequently on 35 paired sera from Patrol IV, 133 from Patrol VII and 50 from Patrol VII-H (i.e. from the tender crew, collected simultaneously with Patrol VII). With the exception of three men on Patrol VII who showed four-fold titre rises to Influenza C, no diagnostic increases were noted with antigens to Influenza A, B and C; Parainfluenza 1, 2, and 3; Herpesvirus; Adenovirus; Mumps; Respiratory Syncytial virus; or to *M. pneumoniae.* In short, the etiology of the patrol IV outbreak was not established, and none of these viruses could be implicated in the mild URI syndromes observed in the other patrols.

6. *Etiological studies: viral isolations.* Approximately 1200 oropharyngeal gargle or throat swab specimens from crew members with or without elicited respiratory symptoms during Patrols V, VI and VII were passed three times in at least two human diploid cell cultures, and usually also in one to three heteroploid cell lines. Smaller numbers of specimens were collected by collaborating medical officers during Phase III, Patrols XI, XIV, XVI and XVIII, and also by one of us (WEB) stationed aboard the submarine tender during the first several weeks after the start of Patrol VII. Specimens from Phase III patrols and from the tender for the most part were obtained from persons with manifest infection. Those obtained from Patrols V, VI and VII were collected at intervals throughout a patrol from nominally equal groups of men with or without symptoms.

Table 8 lists the 35 viral isolations obtained from these specimens, together with the numbers of crew members freshly infected, as judged by four-fold increase in paired serum neutralization titres. No isolations were made after the second week of a patrol with the exception of one Herpesvirus in the seventh week and two Rhinoviruses in the eighth week; the latter, however, were obtained during a period of reinfection after the submarine had entered harbour for two days during the sixth week of the patrol. Seven additional Herpesvirus, one Rhinovirus and two Respiratory Syncytial virus isolations were made from specimens collected early in a patrol; two additional Herpesvirus, two Coxsackie virus, three Echo 28 and two Rhinovirus isolations were made from specimens collected during refit; and eight Rhinovirus, four

Herpesvirus and one Adenovirus isolations were obtained from the tender crew studies in parallel with Patrol VII.

Table 9 summarizes results from serum neutralization surveys with the isolates: in some cases on sera from the entire crew, in others on sera from small test groups or from symptomatic crew members only. In almost every instance a diagnostic serum neutralization titre increase was obtained with homologous paired sera from the crew member providing the viral isolate. The notable exception occurred with the two Respiratory Syncytial isolates (Table 8). However, no evidence was obtained to suggest that any of the isolates was transmitted to any substantial number of a crew, or that any could be implicated in the large numbers of URI present immediately prior to or in the first weeks after deployment to sea. In the seven instances in Table 9 where the entire crew was surveyed, little or no antibody was found in one-third or more of each crew, suggesting the presence of substantial numbers of susceptibles in a crew. There was no indication that the antibody levels in a crew affected transmission of infection.

Further examination was made of the transmission of two Rhinovirus isolates (Table 10), one obtained from a crew member early in Patrol VII, the other, at approximately the same time, from a crew member of the tender, or supply ship, that had just completed re-equipping the submarine. These were titred against all paired sera available from each of the Patrols IV through VII and VII-H (Tender), covering the period of possible exposure of the Blue and Gold crews of this submarine during the preceding ten months. It should be remembered that roughly 25 per cent of each crew represented new men not present on a preceding patrol.

Approximately 80 to 85 per cent of Patrol VII and VII-H crews had little or no antibody, and 6 to 12 per cent had high ($> 1/64$) antibody titres to the Tender Rhinovirus isolate which at that time was an apparently new serotype in that it was not neutralized by some 51 type-specific Rhinovirus antisera. In contrast, approximately 40 and 50 per cent of each crew had low ($< 1/16$) titres, and 50 and 30 per cent respectively had high titres to the submarine Rhinovirus isolate, which was serologically identical with the English serotype B632. Serological evidence of infection with both isolates was detected in 4 per cent of both the submarine and the tender crews, but not in earlier patrols. Although the numbers are too small to be conclusive, there was no indication that the submarine environment or the differing proportions of apparently susceptible crew members (or differing antigenicity of the isolates) affected the transmission of these Rhinoviruses.

Inspection of elicited and check-off URI symptom records from the patrols listed in Table 8 indicated that, with the exception of six Herpesvirus and one Rhinovirus isolates, no viruses were isolated from control groups in the absence of manifest URI symptomatology. It is perhaps noteworthy that the overall isolation rate from specimens collected during Patrols IV–VII, from frozen oropharyngeal gargles, was 4·0 per cent (8/202); for swabs rinsed in broth which were then frozen, 4·0 per cent (8/207); for swabs rinsed in previously collected gargle which were then frozen,

5·0 per cent (4/79). Aboard the tender, the isolation rate from oropharyngeal gargle samples was 11·1 per cent (5/45) and for swabs in broth, subsequently frozen, 14·6 per cent (7/48). These rates were not improved by the initial attempts to isolate specimens by means of immediate inoculation into tissue cells taken aboard the submarine and tender during Patrol VII and VII-H; isolation rates for oropharyngeal and swab samples in broth immediately inoculated were 2·1 per cent (1/48) and 2·7 per cent (2/73).

These rates reflect the overall situations encountered by three investigators working under very disparate conditions and are cited primarily to provide for the reader a grossly qualitative measure of isolation efficiency in this operation. Moreover, one might expect that hand-carrying unrefrigerated, preformed tissue cell monolayers over 6000 miles, on and off planes, and in air temperatures ranging from 60°F to in excess of 100°F, would preclude any cellular survival at all. However, such preparations, without further passage or treatment, permitted four isolations, respectively of two Respiratory Syncytial viruses, an Adeno and a Rhinovirus.

Several hundred air samples from three patrols, collected during periods of inboard venting of sanitary tanks, and representing the air washings of approximately 300 $ft.^3$ of submarine air, were combined and concentrated in the cold; no viruses were isolated.

7. *Crew characteristics associated with URI.* A study was made of medical histories, familial and other personal characteristics of crew members on Patrols V through VIII in an attempt to establish attributes associated with an increase in respiratory infections. The assumption was made that the frequency of elicited symptoms during a patrol provided an index measuring resistance to respiratory infection which could be related to crew characteristics obtained from individual questionnaires completed by each crew member. Additional data were obtained from crew members' medical records.

Information on symptoms of respiratory infections was taken from the symptom check-off lists completed at sea by the entire crew, approximately every five days. In accordance with the number of positive recordings for URI symptoms, a crew was divided into 'high URI' and 'low URI' halves. With the assistance of a computer, 27 crew characteristics were tabulated according to low or high URI status during each patrol at sea, and for all five patrols combined. Chi-square analysis for independence of each variable was determined at the 5 per cent and 1 per cent levels of significance. Table 11 summarizes the principal findings from the combined patrol data.

Region of origin, residence area, marital state, previous medical history other than URI, and certain events during layover periods between patrols showed no relation to URI recorded on patrol. Characteristics which were associated with increased URI primarily fell into three groupings:

1. Factors associated with increased age and responsibilities, both in the service and at home (officers were excluded by insufficient numbers).

2. A subjective impression or history of recent or substantial URI difficulties.
3. Bunk location on patrol: crew members occupying one group of 17 bunks had significantly more URI, not alone in Patrols V through VIII, but also in two other submarines studied in Patrols II and XI. The reason for this difference is currently not known.

Another unexpected association with increased URI was that of CPO rating or of age over 35 years. This would appear to be in contrast to the observation from family studies of respiratory infection that the man of the household has less infection after age 35. This age group, of course, contains the larger number of men with more years of service; it has greater responsibilities in the Navy and in the home; it has a larger number of primary-school-age children. In the family studies, the school-age child is a major source of infection introduced to the family unit. All these factors presumably contribute to increased stress on this older age group.

The association of past URI (recollected) with increased URI on patrol is of added interest in that 'chronic' URI recorders on patrol, that is men who reported positive for URI symptoms at all or all but one recording, were disproportionately numerous (6/29 vs. 5/116, $p<0{\cdot}01$) among crew members developing the mid-patrol syndrome observed in Patrol IV. These observations suggest that, while too restrictive for application to the Polaris program, diligent examination of URI history may be pertinent in screening candidates for prolonged closed cabin voyages.

8. *Polaris submarine Sick Bay visits, 1967–1968.* An important question is, how representative are the observations we have reported for the submarines we have studied, of Polaris submarines in general? This can be answered only indirectly by inspection of the medical reports for each patrol by the medical officer aboard. Such reports vary widely in detail and comprehensiveness according to the predilection of individual medical officers, and are directed primarily towards the recording of illness and accident which could affect the success of the patrol, rather than the burden of minor illness which energetic young men may accept as a normal part of their military career; nevertheless, we have examined in excess of 100 such reports from a 12-month period in 1967–1968, and tabulated data pertaining to illness resulting in Sick Bay visits (Tables 12–16).

Table 12 categorizes all Sick Bay diagnoses into seven groupings, illustrating that URI and gastrointestinal disturbances together represent 60 per cent during refit and 77 per cent during patrol of all medical problems requiring treatment. Whereas there appears to be little difference in the combined rates for diagnoses not attributable to URI and GI disturbances with respect to the four harbors from which Polaris submarines sail, this is not the case for URI and GI infections. We suggest that the added stress of overseas flight to a different microbial and climatic environment, and particularly the latter in the case of Holy Loch, contributes to the increased Sick Bay visits recorded for the overseas harbors. It should be noted in passing that exceedingly few of the Sick Bay reports refer to illness so severe as to be incapacitating.

Tables 13 and 14 classify those patrols on which there occurred an URI or GI outbreak with respect to the season of the year in which the submarine was at sea. The data are too scant for generalization on the basis of one year alone, but it is apparent that a narrow seasonal grouping of outbreaks did not occur and should not be expected with the possible exception of URI at Rota and Charleston.

Tables 15 and 16 regroup patrol URI and GI data to illustrate patrol weeks in which outbreaks occurred. Where a double number is listed (1+24), the first numeral indicates the number of Sick Bay visits resulting in work days lost. A bold number denotes the week in which an outbreak attained peak incidence.

Gastrointestinal disturbances apparently may be expected to occur at any time during the patrol cycle. Since the average number of GI Sick Bay visits was seven for patrols not listed on this table, it appears that several patrols were affected by continuing or repeated infections in low numbers throughout the entire refit and patrol at sea. The reports do not clearly indicate how many men were affected by these minor illnesses.

In contrast, URI outbreaks predominantly occurred during the refit period and, to a lesser extent, in the first two weeks at sea. It should be noted that one outbreak apparently occurred as late as the sixth week at sea; it is not known whether this patrol was earlier broken by a few days in harbor.

Discussion

These studies had three primary objectives. The first was to determine microbiological hazards that afford a substantial immediate threat to the mission and welfare of Polaris submarine crews. We found none. The second was to evaluate the microbiological ecology of the Polaris submarine environment with the objective of improving the habitability of such vehicles. We have found several operational procedures in which change would unquestionably enhance comfort and morale. Such change may not be compatible with the priority and demands of the Polaris mission as presently constituted, but it could be vital to the success of longer-term sojourns beneath the sea or in space. Finally, we have sought to elucidate the epidemiology of respiratory infections, particularly as it is manifested in the closed cabin environment.

The Polaris submarine is not a research tool that can be manipulated to suit the needs of an experimenter. It is not manned by military recruits, the subjects of many URI investigations, but by seasoned volunteer officers and men, who, subjectively selected by factors of individual motivation and objectively screened for psychological stability and physical well-being, are not representative of military populations, much less of civilian family or industrial groups. All this has limited the methods we have employed and the generalizations that we might wish to make.

We know too little concerning the efficiency of the sampling, transport, and cultural procedures we have employed, but we know enough to state that a positive finding generally should be respected but considered minimal; a negative, unsupported by corroborating data, is suspect. From this viewpoint, we have indicated that air microbial sanitation in the Polaris submarine, within the limit of our technics, generally appears to be excellent, and certainly we have not seen any infectious processes to contravene this conclusion. Yet, we have demonstrated a potential in Patrols I through III for periodic massive contamination of the air. We attribute this to the sanitary system of the submarine, but were unable directly to recover high concentrations or organisms during its operation. Nevertheless, if the low levels of staphylococci, coliforms and enterococci recovered from the air can be considered representative of continuing air contamination, one may calculate that, in the course of a patrol, crew members inhale at least 900 staphylococci and coliforms, and 18,000 enterococci; periodically in Patrols II and III dosages were 1000-fold greater.

We also know too little concerning the antigenic impact or infectious response in men continuously exposed by inhalation to very low microbial dosages. Certainly, in Patrol II there was no detected effect of the crew members' harborage of staphylococcal strains of the same phage types found in the air. But neither was there present a strain capable of inducing manifest disease.

Some years ago L. J. Goldberg of our Laboratory demonstrated that inhalatory exposure of several hundred mice to incrementally administered dosages of *Klebsiella pneumoniae*, such that a total lethal dosage was received over a period of some days, resulted in the same number of fatalities in the herd, within statistical limits, as was obtained by short-term inhalation of the same dosage. However, the response curve became skewed towards increased survival when administration of small dosages was prolonged beyond 30 days; this was thought to be the effect of an antigenic response in some members of the herd (13).

If a similar process is operative in the Polaris submarine environment, as we assume, then prolonged low exposure to airborne micro-organisms of low virulence or transmissibility (including poor capacity for airborne survival) would contribute towards development of resistance in the crew to infection by those organisms. By the same token, introduction or activation of a virulent strain, or one with a greater capacity for contact or airborne transmission and survival, may be expected to produce an outbreak of infection. No such outbreak with bacterial infections occurred in the patrols we have studied; however, we consider that the potential exists.

Introduction of new viral strains is believed to be the cause of recrudescence of URI symptoms in the crews of Patrols III and VI following brief return to harbor late in a patrol. To a degree, this interpretation is supported by our isolation from the Patrol VI crew of two Rhinoviruses at this time, the only instance in which any viral isolation was made later than the second week of a patrol.

Activation of an unidentified virus present in the crew is a plausible explanation of the mid-patrol outbreak that occurred in the third week at sea on Patrol IV (14). That the infection differed from the customary mild URI early in patrol was indicated not only by the difference in symptomatology we have described, but also by the observation that it affected equal numbers of men who had or had not experienced mild URI infections earlier in the patrol. That is was spread primarily by close contact was demonstrated by our observation of a statistically significant ($p < 0.05$) increase of secondary cases occupying the same tier of bunks with primary cases, as opposed to other berthing tiers.

Yet, neither in the above outbreaks nor with some 25 other viral isolations did we obtain serological evidence linking the isolates etiologically with the widespread mild URI observed. The failure does not seem to us attributable to technic or to time of collection of paired sera, since for all but four isolates, diagnostic titre increase was demonstrated. We repeated a few of the Rhinovirus serum neutralization survey tests with reduced virus (3 and 10 TCD_{50}) without beneficial results, but we have not tried the microplaque reduction technic found by Taylor-Robinson and Tyrrell (15) to have greater sensitivity for detection of Rhinovirus antibody. Other deleterious factors were that we sought isolates primarily during the decline of the URI epidemic, unknown portions of our specimens may have been mishandled during transport, the standard isolation systems employed may be inadequate for isolation of many viruses contributing to mild URI (4, 16) and, finally, such viruses may be poorly antigenic. Whatever the reasons, our data indicate that a substantial amount of minor respiratory infection of unknown etiology afflicts Polaris crews during refit and the early weeks of many patrols at sea, and that, on occasion, there is a more severe respiratory disease outbreak in mid-patrol. It is noteworthy that one such severe outbreak was shown by retrospective serology to have been caused by *Mycoplasma pneumoniae* (17).

Two earlier studies have been reported on microbial air contamination in submarines. Boyden in 1962 studied air contaminants by means of settling plates containing blood agar or endo agar, which were opened in various compartments of a Polaris submarine before and during a patrol at sea (18). His air counts were also low, but cannot be compared accurately with ours because of the differences in collection procedures. He also recovered coliform organisms and associated levels of air contamination with the degree of crew activity in a given compartment.

Boyden also reports weekly incidence of URI based upon Sick Bay visits; his curve resembles that of our Patrols II and III, peaking in third week at sea after substantial infection during a two-week refit period.

In World War II, Ellis and Raymond did a detailed study of bacterial air hygiene in a cruiser and a diesel-powered submarine, using Bourdillon slit samplers for collection of bacteria primarily on blood agar plates as they were interested in contamination by hemolytic organisms (19). They included in their report an interesting summary of respiratory disease outbreaks on British warships in both World Wars, noting:

> 'Men suffering with minor respiratory infections frequently fail to report to the Sick Bay, thus the true incidence of these infections will not be shown by an

examination of sickness rates, but few will refute the claim that these epidemics exercise an adverse effect on general efficiency.'

and,

'Even when the fact that submarine crews are carefully selected is taken into account, the low incidence of respiratory tract infections has frequently been discerned by medical officers of submarine flotillas, some of whom had anticipated that the conditions under which these small but compact communities were living during the long diving patrols . . . would be conducive to the spread of infectious disease.'

Their words are equally descriptive of present studies.

Similarly, a summation of McLean's findings 50 years ago (6), on the Antarctic epidemiology of respiratory infections in isolated populations, was but little extended in the latest study we have seen of an Antarctic community (8). McLean found:

1. Men underwent an initial period of susceptibility to respiratory infections.
2. After recovery from initial infections, men apparently entered a period of immunity to endogenous respiratory infection.
3. Following contact with relief personnel from outside, the men again experienced respiratory infections; this time the infections were of some severity.

Cameron and Moore came to similar conclusions expressed in the current terminology of virus infections and, using isolation and serological technics remarkably similar to those we have employed in these studies, they, too, were unable in the laboratory to establish the etiology of the respiratory infections they encountered (8). We can only conclude that the solution to this problem must await development of better technics, perhaps including the use of organ cultures similar to those in which Tyrrell and his associates have detected previously unsuspected Rhinoviruses (20, 21).

The control of URI in human populations has made notably little progress, in large measure because of the ubiquity of respiratory viruses that confer little or no cross-protection. Yet, in the past 40 years of experimental and clinical epidemiological investigations since Webster's (22) and Greenwood's (23) classical experiments with controlled infections in mouse populations, many principles have been established which we can utilize to mitigate or enhance herd infection or immunity.

A review of the Polaris environment and operational cycle earlier described (pp. 9–10) suggests that inadvertently these principles are both honoured and violated, due to the exigencies of the Polaris mission. Let us illustrate by redescribing the Polaris cycle in somewhat different terms:

A population of 130 to 140 young men, selected by physical examination and medical records for their natural resistance to infection, are brought into close proximity over several days prior to abrupt transfer to and temporary housing in a new climate which is frequently colder and damper. This population has recently lost about 15 to 25 per cent of its seasoned members, with replacement by an equal number of younger, inexperienced immigrants presumably with a different recent infection history. After several days of crowded discomfort, the population is housed but not confined in a constant climate chamber maintained at nominally 70°F, and 50 per cent relative humidity, whatever the external meteorological conditions. The population is now stressed for four weeks by unusual fatigue, and possibly by psychological factors deriving from recent separation from family and family problems. The last might particularly apply to men with several young children, one or more of whom may have been ill at departure, providing an input of infectious agents prevalent in the home community. During this four-week period there is daily ingress and egress of temporary immigrants. Any member of the population who becomes seriously ill is immediately removed; all others, after developing some degree of equilibrium and hard immunity towards local infectious agents, are sealed off in the constant climate chamber and observed for two months. In the course of several more weeks acquired resistance diminishes infection to a negligible level which usually is maintained so long as isolation is preserved.

Until such time as means are available to enhance nonspecific resistance to infection, planners of prolonged space or submerged voyages in closed cabin vehicles should seek to develop departure regimens in keeping with known principles affecting herd infection and immunity.

Acknowledgements

This project was sponsored by the U.S. Navy Special Projects Office and Director of NBL. The cooperation is acknowledged of Naval Medical Officers on Polaris Patrols, the staff at NBL and 6th Army Medical laboratories in the collection and processing of specimens, and preparing statistical data.

REFERENCES

1. Calloway, D. H., Editor (1967). *Human Ecology in Space Flight*, Vol. II, New York Academy of Sciences, New York, 298 pp.
2. Homna, M. and Crosby, H. J., Editors (1963). *A Symposium on Toxicity in the Closed Ecological System*, Lockheed Missile and Space Co., Palo Alto, Calif.
3. Kinsey, J. L. (1960). Some toxicological hazards in submarines, *Fed. Proc.*, **19,** 36–39.
4. Tyrrell, D. A. J. (1965). Epidemiology, pp. 19–44. In *Common Colds and Related Diseases*, William and Wilkins Co., Baltimore.

5. Symposium on medical problems presented by the Tristan da Cunha community (1963). *Trans. R. Soc. Trop. Med. Hyg.*, **57,** 19.
6. MacLean, A. L. (1919). Bacteriological and other researches, *Scient. Rep. Australas. Antarct. Exped.*, **7** (part 4), 127.
7. Sladen, W. J. L. (1965). Staphylococci in noses and streptococci in throats of isolated and semi-isolated Antarctic communities, *J. Hyg., Camb.*, **63,** 105–116.
8. Cameron, A. S. and Moore, B. W. (1968). The epidemiology of respiratory infection in an isolated Antarctic community, *J. Hyg., Camb.*, **66,** 427–437.
9. Sherman, J. M., Mauer, J. C., and Stark, P. (1937). Streptococcus fecalis, *J. Bacteriol.*, **33,** 275–282.
10. Finegold, S. M. and Sweeny, E. E. (1961). New selective and differential medium for coagulase-positive staphylococci allowing rapid growth and strain differentiation, *J. Bacteriol.*, **81,** 636–664.
11. Blair, J. E. and Williams, R. E. O. (1961). Phage typing of Staphylococci, *Bull. World Health Org.*, **24,** 771–784.
12. Deig, E. F. and Watkins, H. M. S. Manuscript in preparation.
13. Goldberg, L. J. (1952). Mortality in a herd of mice as a function of time: effect of an immune response. *Sixth Technical Progress Report*, Naval Biological Laboratory, University of California, Berkeley.
14. Andrewes, C. H. (1964). The complex epidemiology of respiratory virus infections, *Science*, **146,** 1274–1277.
15. Taylor-Robinson, D. and Tyrrell, D. A. J. (1962). Serological studies on some viruses isolated from common colds (rhinoviruses), *Brit. J. Exptl. Path.*, **43,** 264–275.
16. Andrewes, C. (1965). *The Common Cold*, Weidenfeld and Nicolson, London, 187 pp.
17. Sawyer, R. and Sommerville, R. G. (1966). An outbreak of *Mycoplasma pneumoniae* infection in a nuclear submarine, *J. Am. Med. Assoc.*, 1965, 174–175.
18. Boyden, D. G. (1962). The bacterial flora in fleet ballistic missile submarines during prolonged submergence, *U.S. Naval Medical Research Laboratory Report*, No. 386, Vol. 21 (17).
19. Ellis, F. P. and Raymond, W. F. (1948). VIII. Air hygiene in H.M. ships under wartime conditions. 40. Bacterial content of the air in a cruiser and a submarine. In Bourdillon, R. B., Lidwell, O. M. and Lovelock, J. E., *Studies in Air Hygiene*, Medical Research Council Special Report Series No. 262, London.
20. Hoorn, B. and Tyrrell, D. A. J. (1965). On the growth of certain 'newer' respiratory viruses in organ culture. *Brit. J. Exptl. Path.*, **46,** 109.
21. Tyrrell, D. A. J. and Bynoe, M. L. (1965). Cultivation of a common cold virus in organ culture. *Brit. Med. J.*, **i,** 76.
22. Webster, L. T. (1933). Summarized by Wilson, G. S. and Miles, A. A. (1955):

Topley and Wilson, *Principles of Bacteriology and Immunity*, Fourth ed., Chapter 56, Williams and Wilkins Co., Baltimore.

23. Greenwood, M., Hill, A. B., Topley, W. W. C. and Wilson, J. (1936). *Experimental Epidemiology*, Special Report Series No. 209, Medical Research Council, London.

TABLE 1

Preparations tested for storage of micro-organisms at reduced temperatures

Agar	0·2%
Ascorbic acid	0·01%
Casein, isoelectric	2%
Casein, purified	2%
Casein, technical	2%
Caseinate, sodium	2%
Gelatin	0·5%, 2%[a], 5%
Gelatin (2%)+ skim milk (5%)[a]	
Gelatin phosphate diluent	
Horse serum	10%, 20%[a], 50%[a]
Horse serum (25%)+ skim milk (5%)	
Lactalbumin	2%
Lactalbumin hydrolysate	2%
Nutrient broth	
Parenamine	1·5%, 3%
Peptonized milk	3%[a]
Proteose peptone	0·2%
Saline	
Skim milk	0·5%, 5%, 10%
Skim milk (5%)+ gelatin (2%)+ starch (0·05%, 0·1%, 0·5%, 1%)	
Skim milk (5%)+ gelatin (2%)+ horse serum (10%)+ starch (0·05%, 0·1%, 0·5%, 1%)	
Stuart's amino acid	1·5%, 3%
Stuart's medium	
Thioglycollate	

[a] Preparations giving 90 to 100% recovery after 6 weeks at −70°C.

TABLE 2

PATROL II: MULTIPLE DETECTION[1] OF HIGH CONCENTRATIONS OF AIRBORNE MICROORGANISMS[2]

DAY	TIME	STATIONS 1	2	3	4	5	6	7	8	TOTAL[3]
9	0900				X				X	2
12	0900			X			X	X	X	4
14	0300			X	X	X	X			6
	0900				X	X				
15	0900	X	X	X		X	X			5
17	0900				X			X		2
23	0900				X					2
	2100				X					
30	2100		X						X	2
31	0300	X	X		X				X	5
	1500					X				
34	0900							X		2
	2100					X				
35	0300						X			2
	0900			X						
39	0300			X		X				4
	0900								X	
	1500				X					

(1) Single detections were made 14 times
(2) High concentration: $>10^3$/liter air
(3) Total detections: 50 on 25/47 days

TABLE 3

Patrol I: Bacteria in air and crew samples

Sample	Air		Skin		Nasal		Oropharyngeal	
Patrol	2	3	2	3	2	3	2	3
Staphylococci (Coag +)	5/368 [a]	4/61	48/510	15/210	101/465	28/203	9/481	8/186
Enterococci	117/368	4/61	30/510	10/210	9/289	5/203	8/74	5/19
Coliforms	7/368	0/61	25/510	10/210	40/289	27/203	0/79	0/31

[a] 5/368= No. positive/total samples

TABLE 4

Bacteria in submarine air: Frequency and magnitude of microbial counts.

$(B)^a$	Patrol II				Patrol IV				Patrol V				Patrol VI				Patrol VIII			
	S_4[b]	%	S_7	%	S_4	%	S_7	%	S_4	%	S_7	%	S_4	%	S_7	%	S_4	%	S_7	%
$\geq 10^2$	15	(9)	11	(6)	0	(0)	4	(3)	0	(0)	0	(0)	0	(0)	0	(0)	0	(0)	0	(0)
$\geq 10^1, < 10^2$	17	(10)	10	(6)	9	(8)	4	(3)	4	(3)	4	(3)	1	(1)	0	(0)	5	(2)	1	(0)
$\geq 10^0, < 10^1$	63	(36)	18	(10)	59	(52)	26	(21)	33	(32)	18	(17)	52	(31)	13	(8)	43	(19)	29	(13)
$\geq 10^{-1}, < 10^0$	63	(36)	42	(24)	31	(27)	38	(30)	49	(48)	43	(41)	84	(51)	69	(41)	96	(43)	113	(48)
$< 10^{-1}$	16	(9)	95	(54)	15	(13)	53	(42)	38	(37)	38	(36)	29	(17)	87	(51)	81	(36)	83	(37)
Totals	174		176		114		125		102		105		166		169		225		226	
Av/day	3.7		3.7		2.5		2.8		1.6		1.7		3.0		3.1		4.0		4.0	

[a] (B): Total viable bacteria per liter air; $< 10^{-1}$ = no recovery.

[b] S_4: Crew's lavatory; S_7: passageway near mess compartment.

TABLE 5

Patrol II: distribution of staphylococcal phage types in test group[a]

Phage type[b]	*Carriers*[c]	*Transient*	*Acquired new type*	*Lost type*
52/52A/80	3	1	2	1
52/79	2	0	0	0
3A/3C/71	1	0	1	0
6/42E/53/83A	2	4	0	0
Total	8	5	3	1

[a] No coagulase-positive staphylococci recovered from 13/30 men tested.
[b] Abbreviated designations.
[c] Crew sampled 17 times (2X/week) by nasal or skin swab. Carrier: positive intermittently or regularly throughout patrol.

TABLE 6

Patrol IV: distribution of staphylococcal phage types in crew[a]

	Phage type[b]	*Carriers*[c] *(6–8 pos.)*	*Irregular harborage (3–5 pos.)*	*Transient (1–2 pos.)*	*Acquired new type*[d]	*Lost type*
A	52/52A/80	8	12	1	1 TC	1
B	52A/79	2	1	1	1 TE	0
C	3A/3C/55/71	4	3	3	1 TA	1
D	42E/47/54/83A	13	2	3	1 D+1 TE	1
E	187	4	0	1	0	0
F	Not typable	0	5	4	1 TA	0
	Total	31	23	13	1 D+5 T	3

[a] No coagulase-positive staphylococci recovered from 75/145 men tested.
[b] Abbreviated designations.
[c] Crew sampled eight times (1X/week) by nasal swab.
[d] TC: temporary acquisition of type indicated. One man acquired and retained type D.

TABLE 7

Patrol IV: micro-complement fixation titres

Antigen used	*Total sera*	*Number of pre and post sera showing titre*					
		<8	*8*	*16*	*32*	*64*	*128*
Influenza A	170	9	32	58	59	8	4
Influenza B	170	35	53	56	21	5	0
Influenza C	170	21	34	69	37	9	0
Adenovirus	170	12	16	34	67	26	15
Herpesvirus	170	49	36	30	44	10	1
Mumps	170	38	61	44	17	9	1
Poliovirus I	170	100	36	28	6	0	0
Poliovirus II	170	126	26	9	9	0	0
Poliovirus III	170	70	52	37	6	4	1
Parainfluenza 1	170	70	55	37	8	0	0
Parainfluenza 2	94	77	15	2	0	0	0
Parainfluenza 3	170	18	46	63	39	3	1
Respiratory Syncytial	170	79	63	27	1	0	0

TABLE 8

Virus isolations from crews of submarines and tender

Patrol	*Virus isolated*	*Type*	*Number isolations*	*Diagnostic serum titre (homologous)*
V	Herpes simplex		1	1/1
VI	M-Rhinovirus	H26	2	2/2
	-Rhinovirus	ND	1	1/1
	Herpes simplex		3	ND
VII	M-Rhinovirus	B632	1	1/1
	-Rhinovirus	ND	1	1/1
	Herpes simplex		3	3/3
	Resp. Syncytial		2	0/2
VII-H	M-Rhinovirus-35T3	ND	3	3/3
(tender)	H-Rhinovirus-101T	NNT	4	1/2
	Herpes simplex		3	ND
	Adenovirus	4	2	1/2
	Coxsackie	B3	1	1/1
XI	Rhinovirus	ND	1	1/1
	Herpes simplex			ND
	Adenovirus	3	1	11/1
XIV	Echovirus	28	3	3/3
	Coxsackie	A21	1	1/1
XVI	Coxsackie	B5	1	1/1
	Herpes simplex		1	ND

ND: Not determined.
NNT: Not neutralized by 51 Rhinovirus type-specific antisera.

TABLE 9

Neutralization of patrol sera by submarine isolates

Patrol antisera		Patrol and virus	Number of sera showing titre					
			Low		High		Diagnostic	
			< 8	< 16	⩾ 16	⩾ 64	rise (⩾ 4X)	Total
IV	PRE[a]	VII 97 P2A Rhino	25	—	—	11	—	132
	PRE	VII-H, 101T Rhino	97	—	12	—	—	131
V	PRE	VII 97P2A Rhino	26	43	—	45	—	147
	POST		—	7	—	14	0	28
V	PRE	VII-H, 101T Rhino	86	98	4	—	—	112
	POST		—	27	—	2	0	30
VI	PP	VI 35T3 Rhino					ND	ND
VI	PP	VI 51P11 Rhino	13	17	—	2	4	26
		25P11 Rhino	—	24	—	3	1	35
VII	PP	VII 101P2B Rhino	8	—	—	4	2	17
		VII-H, 35T3-M Rhino	32	—	—	10	0	42
		VII-H, 3P1B-H Rhino	9	—	—	1	0	10
		VII 35P1A Resp. Syn.	—	28	—	0	0	28
VII-H (tender)	PP	VII-H, 106P Adeno 4	2	—	—	1	1	7
XI	PP	XI 13-3 Adeno 3	46	—	—	44	2	134
		XI 14-1 Rhino	19	—	—	0	2	41
		XIV 28T Echo 28	—	77	—	0	1	116
		XIV 24T Cox. A21	114	—	—	6	0	130
XIV	PP	XIV 28T Echo 28	9	—	—	5	5	19
		XIV 24T Cox. A21	18	—	—	1	1	21
XIV	PP	XIV 134T Cox. B5	16	—	—	2	1	20

[a] PRE: serum collected at beginning of patrol.
POST: serum collected at last week of patrol.
PP: paired sera from symptomatic and/or control groups.

TABLE 10

Serum neutralization titres to two Rhinovirus isolates from Patrol VII submarine and tender crews

Patrol		*Number of sera showing titre*			
Isolate	*Antisera*	*Low* (< 1/16)	*High* (≥ 1/64)	*Diagnostic rise* (≥ 4X)	*Total* (*pairs*)
VII-97P_2A	VII	71 (53%)	34 (26%)	5 (4%)	133
(B632)	VII-H	20 (41%)	25 (51%)	2 (4%)	49
	VI	24 (37%)	21 (32%)	0	65
	V	43 (29%)	52 (35%)	0	147
	IV	39 (30%)	47 (36%)	0	131
VII-H	VII-H	40 (82%)	3 (6%)	2 (4%)	49
101T	VII	114 (84%)	16 (12%)	6 (4%)	136
(tender)	VI	50 (77%)	6 (9%)	0	65
	V	120 (85%)	9 (6%)	0	142
	IV	111 (85%)	0 (0%)	0	141

TABLE 11

Crew characteristics associated with URI[a]

Associated[b]	*Not associated*[c]
Age: < 20 or > 35 years[d]	Region of residence when recruited
Long service in Navy/submarines	Place where trained
Chief petty officer rating[d]	Residence during layover
More or primary school children[d]	Marital state (divorce excluded)
Increased number colds per year[d]	Allergy in self or family
More colds than fellows/family[d]	Past bronchitis, pneumonia, pleurisy
More severe colds than fellows/family[d]	Hospitalization or illness during training
Had cold during layover	Number of type childhood illnesses
Chronic cold symptoms recent months[d]	Recent weight change
History of sinus trouble	Increasing years in Polaris
Smokes	
Bunk location area	

[a] Thirty-seven variable tested. [b] $P < 0{\cdot}05$. [c] $P > 0{\cdot}05$. [d] $P < 0{\cdot}01$.

TABLE 12

Polaris Sick Bay visits, 1967/1968[a]

Sick Bay visit diagnosis	*Total visits, all squadrons*				*Average visits/patrol/week*			
	Refit	*%*	*At sea*	*%*	*14 Holy Loch*	*18 Charleston*	*16 Rota*	*15 Guam*
URI	2891	54	5623	65 }	6·5	3·5	4·0	4·9
GI	343	6	1004	12 }				
INF, oth.								
Allergy								
Injury	2153	40	1960	23	4·5	4·1	4·2	4·5
Dental								
Other								

[a] Table does not include all patrols during 1967–1968.

TABLE 13

Selected patrols: URI Sick Bay visits/patrol by season[a]

Squadron and refit port	Inclusive months, refit and at sea											
	M–J	*J–A*	*J–S*	*A–O*	*S–N*	*O–D*	*N–J*	*D–F*	*J–M*	*F–A*	*M–M*	*A–J*
14, H. Loch				104	164		93		104		119	79
				90	144				90			
				97					147			
16, Rota										115		
										88		
18, Chstn						80	83	85				
15, Guam			112		86				87		100	82
Average APNS[b]	33	39	36	24	20	37	26	38	39	47	30	15

[a] Patrols selected reported URI Sick Bay visits = 80/Patrol.
[b] APNS: all patrols not selected for inclusion in this table.

TABLE 14

Selected patrols: gastrointestinal Sick Bay visits/patrol by season[a]

Squadron and refit port	Inclusive months, refit and at sea											
	M–J	*J–A*	*J–S*	*A–O*	*S–N*	*O–D*	*N–J*	*D–F*	*J–M*	*F–A*	*M–M*	*A–J*
14, H. Loch					21 50	50			18		34	
16, Rota		41		19				53	49			
18, Chstn						43	30 31			65		
15, Guam				22 18		20				18	52	
Average APNS[b]	2	12	7	5	6	8	5	9	5	7	5	4

[a] Patrols selected reported GI Sick Bay visits $\geqslant$ 18/patrol.
[b] APNS: all patrols not selected for inclusion in this table.

TABLE 15

Selected patrols, 1967–1968: gastrointestinal Sick Bay visits/week[a]

		Refit, weeks					*At sea, weeks*										
Squadron	*Season*	*05*	*04*	*03*	*02*	*01*	*1*	*2*	*3*	*4*	*5*	*6*	*7*	*8*	*9*	*10*	*Total*
14	S–N		**10**	1	2	0	0	1	1	1	1	3	0	1			21
	S–N		0	3	1	0	8	2	6	2	7	11	8	2			50
	O–D		(No information)				2	5	4	1+4[b]	8	4	5	**11**	6		50
	J–M		1	0	0	1	0	**10**	1	2	3	0	0	0			18
	M–M		3	1	0	0	6	3	2	3	5	2	9	0			34
16	Ju–A		2	0	1	5	4	1+6	1+5	5	5	5	1+0	0			49
	D–F		6	1	5	2	7	**12**	7	0	1	1+4	3	3			53
	J–M		0	1+0	0	0	0	0	9	**33**	2	1	1+0	2+0			49
18	O–D		0	0	1	2	2	3	6	4	**10**	5	2	3	2	3	43
	N–J		1	4	2	2	3	2	2	1	4	3	1	3	2	0	30
	N–J		**8**	5	3	3	0	1	3	2	0	2	0	3	0	0	31
	F–A		4	6	8	1	**10**	8	7	0	0	6	3	2	4	6	65
15	A–O		7	7	2	1	0	0	1	1	1	1	0	1			22
	A–O	**16**	0	0	1	0	0	0	0	1	0	0	0	0			18
	M–M		0	4	7	5	1	3	4	8	5	6	3	3	3		52

[a] Selected patrols reported Sick Bay visits ⩾ 10 in week or ⩾ 20 in patrol with average number = 38/patrol. Patrols not included reported average of seven Sick Bay visits/patrol.

[b] 1+4: Sick Bay visits with and without sick days off duty.

TABLE 16

Selected patrols, 1967–1968: URI Sick Bay visits/week[a]

Squadron	Season	Refit, weeks				At sea, weeks										Total
		04	03	02	01	1	2	3	4	5	6	7	8	9	10	
14	A–O	2	**27**	**45**	12	9	2	1	3	1	0	2	0			104
	A–O	9	10	8	10	**14**	7	8	3	8	5	3	5			90
	A–O	**33**	1+**24**[b]	11	7	1	3	4	4	3	1	2	3			97
	S–N	1+**118**	**30**	5	4	3	0	0	1	0	1	1	0			164
	S–N	**24**	**15**	9	**15**	10	**18**	11	9	11	11	7	4			144
	N–J	10	3	**15**	6	**14**	6	7	5	1+9	4	8	5			93
	J–M	**25**	**25**	22	13	11	2	1	2	0	2	1	0			104
	J–M	8	8	11	**17**	11	**19**	14	2	0	0	0	0			90
	J–M	**36**	**42**	23	7	4	**20**	11	2	1	1	0	0			147
	M–M	7	9	8	10	**23**	9	14	**22**	8	4	5	0			119
	A–J	0	0	4	12	**28**	**20**	11	1	1	0	1	0			79
16	F–A	4+7	**5+15**	15	**17**	**16**	9	10	3	5	8	1	0			115
	F–A	11	17	**1+22**	8	10	6	3	2	3	1	3	1			88
18	O–D	0	5	7	8	**16**	8	5	5	3	6	5	2	5	5	80
	N–J	**40**	**25**	1+5	0	0	0	2	1	0	1	3	4	0	0	83
	D–F	**20**	13	4	6	3	7	1	2	5	**19**	4	1	0	0	85
15	J–S	5	4	9	5	7	11	**1+18**	**17**	12	13	6	4	0		112
	S–N	**28**	4	**23**	11	6	3	6	2	0	1	1	0	1		86
	J–M	4	**27**	10	5	**17**	4	4	7	3	2	1	1	1		87
	M–M	0	12	**18**	**26**	1	**16**	9	7	5	2	1	3	0		100
	A–J	6	**13**	10	**15**	8	1+5	1+2	3	8	2	3	3	2		82

[a] Selected patrols reported Sick Bay visits $\geqslant 79$ in patrol with average number = 100/patrol. Patrols not included reported average of 32 Sick Bay visits/patrol.

[b] 1+24: Sick Bay visits with and without sick days off duty.

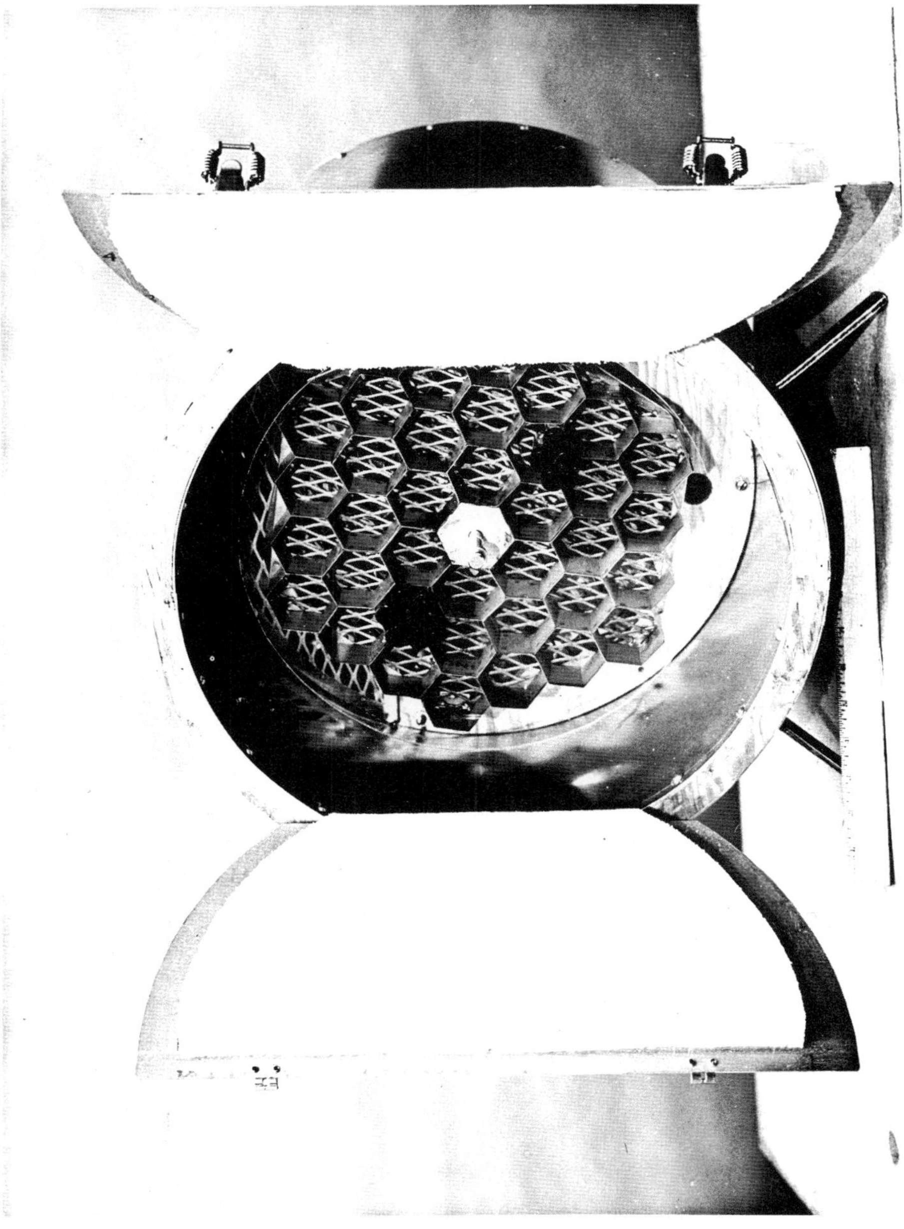

Figure 1. Portable roller drum incubator, front view.

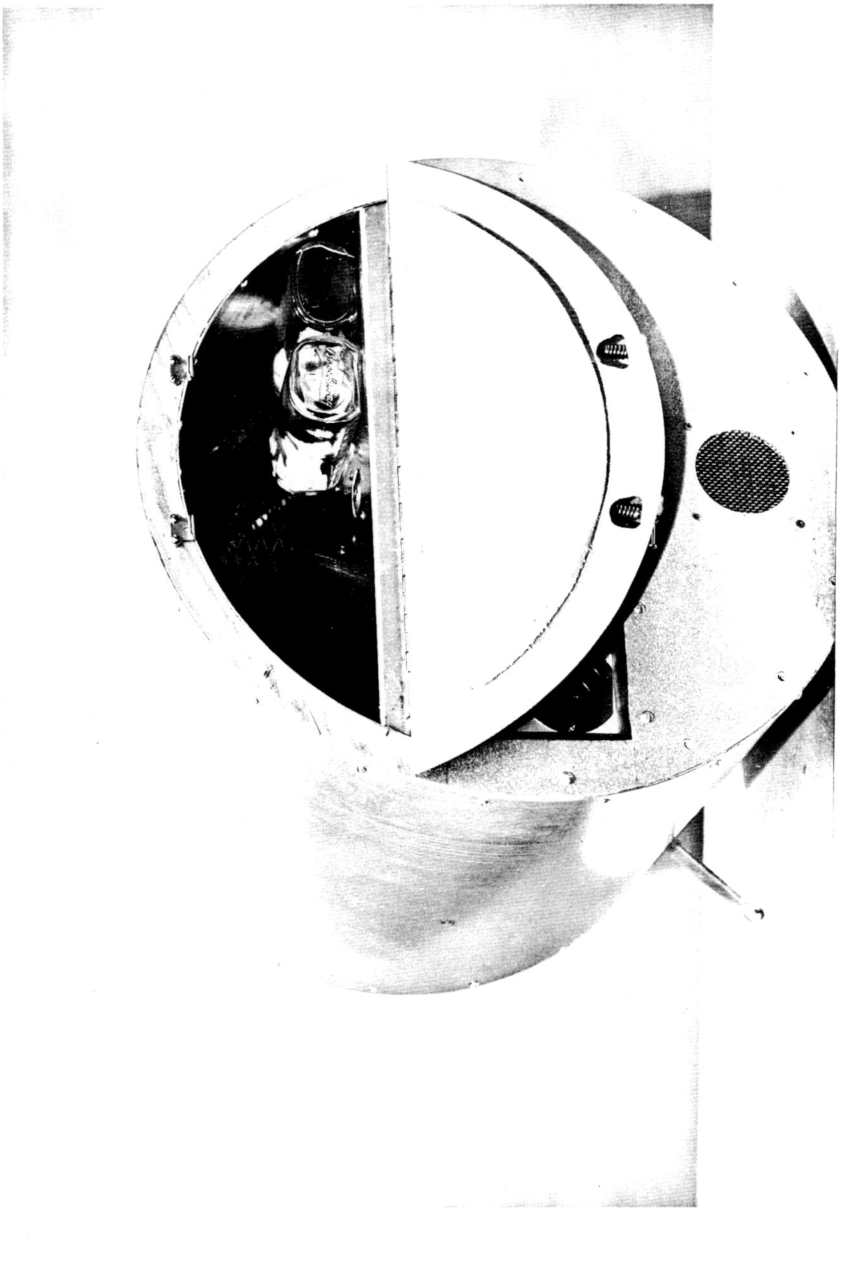

Figure 2. Portable roller drum incubator, rear view.

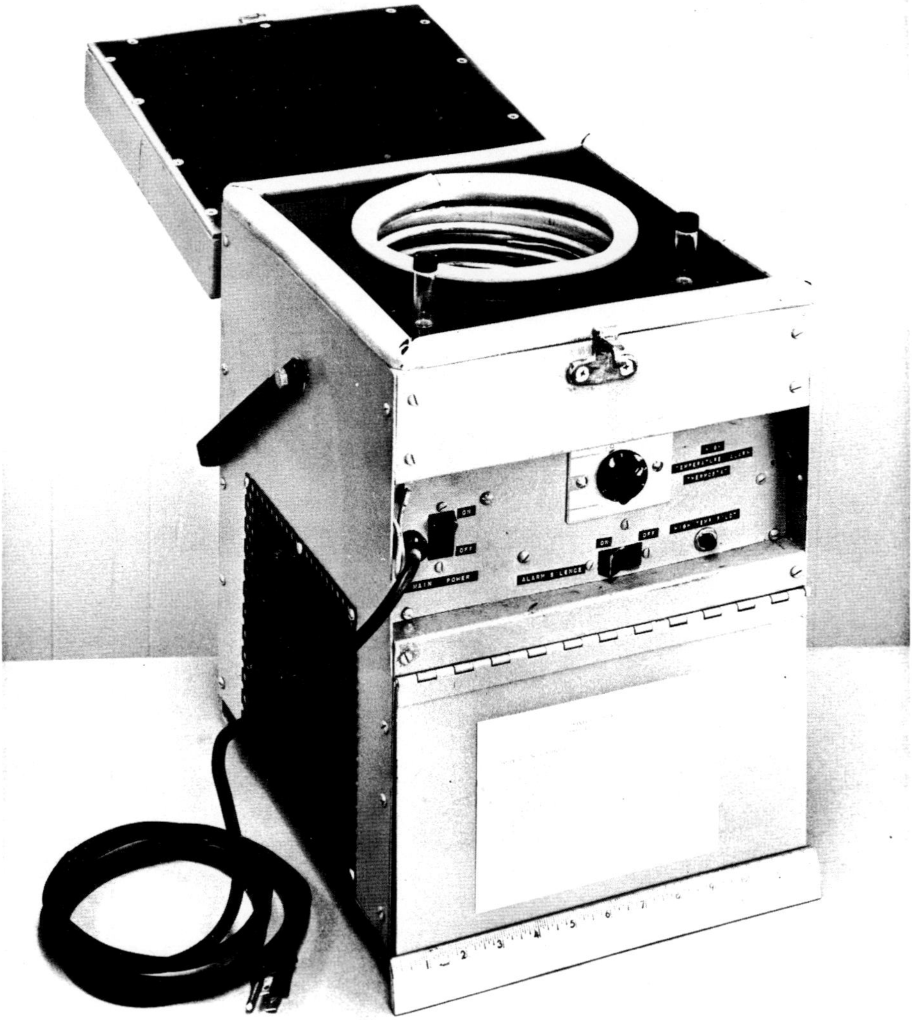

Figure 3. Portable miniature deep freezer.

Figure 4. Air-sampling impinger and pump installation.

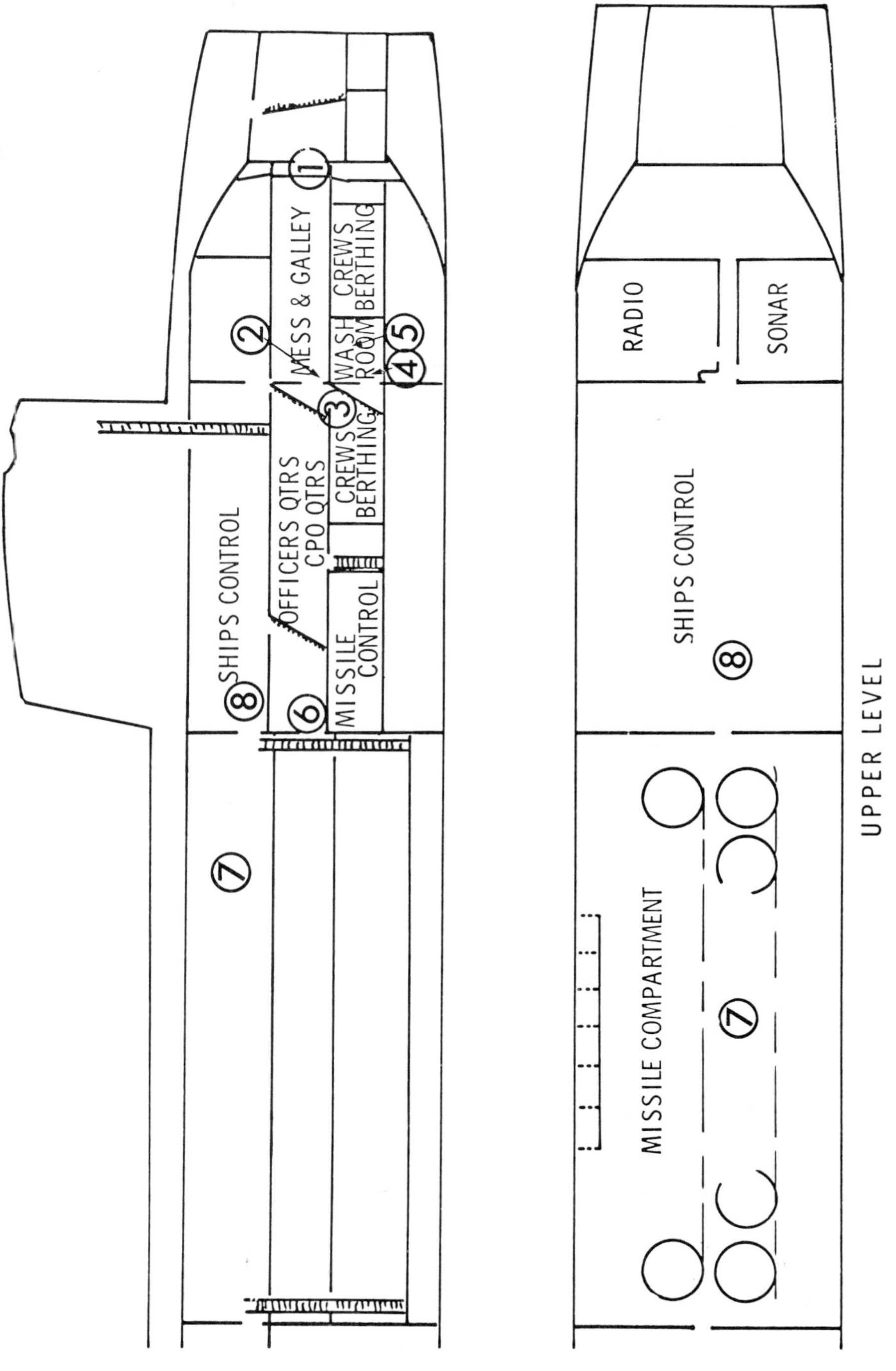
SHIPS CONTROL
OFFICERS QTRS
CPO QTRS
MESS & GALLEY
CREWS BERTHING
WASH ROOM
CREWS BERTHING
MISSILE CONTROL
1
2
3
4
5
6
7
8
RADIO
SONAR
SHIPS CONTROL
8
MISSILE COMPARTMENT
7
UPPER LEVEL

Figure 5

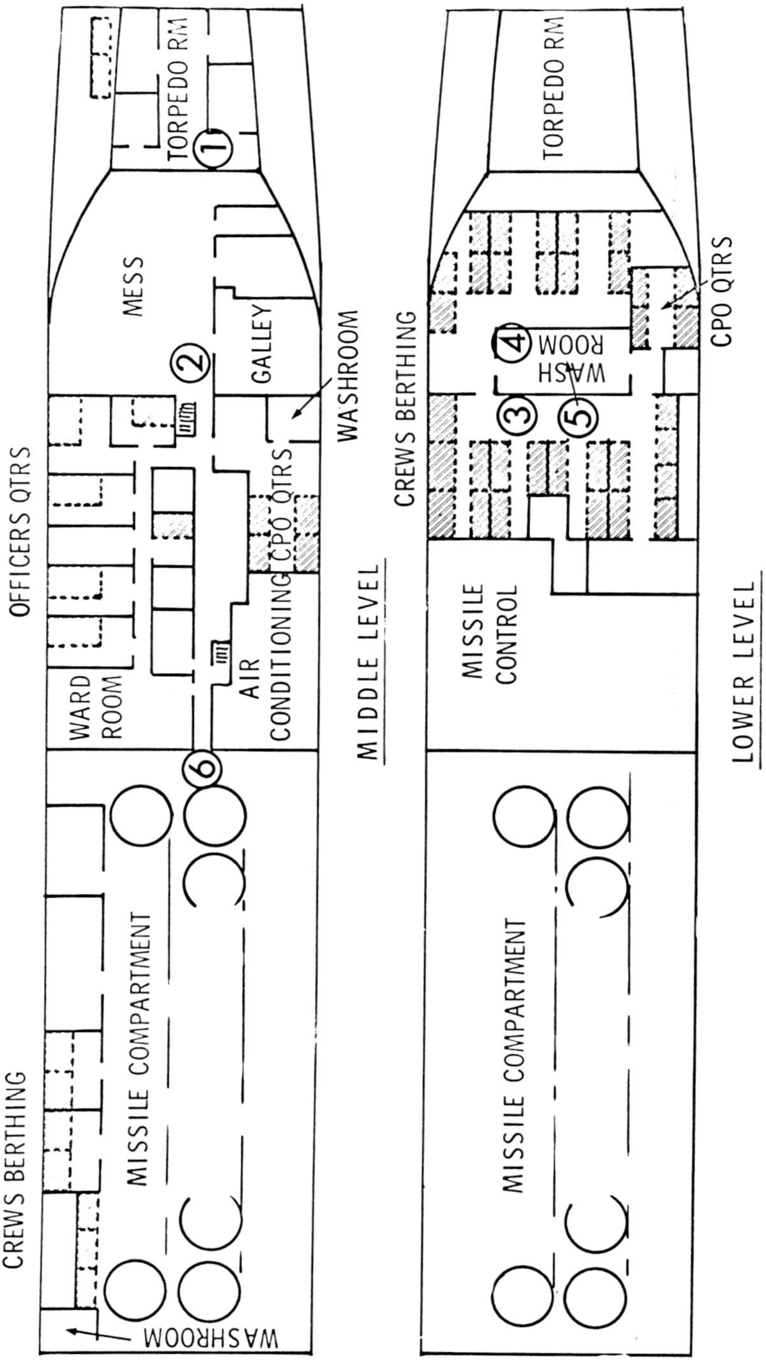
TORPEDO RM
MESS
GALLEY
WASHROOM
OFFICERS QTRS
CPO QTRS
WARD ROOM
AIR CONDITIONING
MIDDLE LEVEL
CREWS BERTHING
MISSILE COMPARTMENT
WASHROOM
TORPEDO RM
CPO QTRS
WASH ROOM
CREWS BERTHING
MISSILE CONTROL
LOWER LEVEL
MISSILE COMPARTMENT
1
2
3
4
5
6

Figure 6

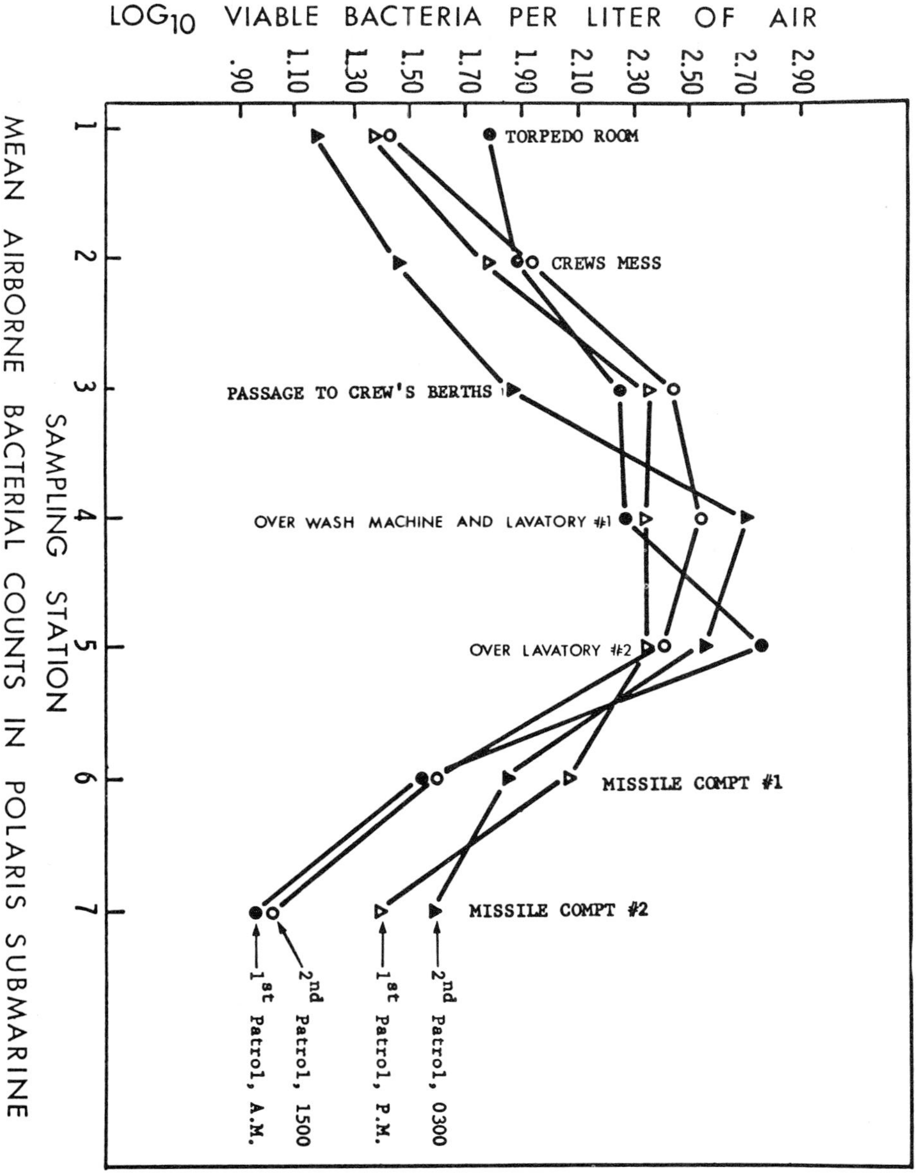
LOG10 VIABLE BACTERIA PER LITER OF AIR
.90
1.10
1.30
1.50
1.70
1.90
2.10
2.30
2.50
2.70
2.90
MEAN AIRBORNE BACTERIAL COUNTS IN POLARIS SUBMARINE
SAMPLING STATION
1
2
3
4
5
6
7
TORPEDO ROOM
CREWS MESS
PASSAGE TO CREW'S BERTHS
OVER WASH MACHINE AND LAVATORY #1
OVER LAVATORY #2
MISSILE COMPT #1
MISSILE COMPT #2
1st Patrol, A.M.
2nd Patrol, 1500
1st Patrol, P.M.
2nd Patrol, 0300

Figure 7

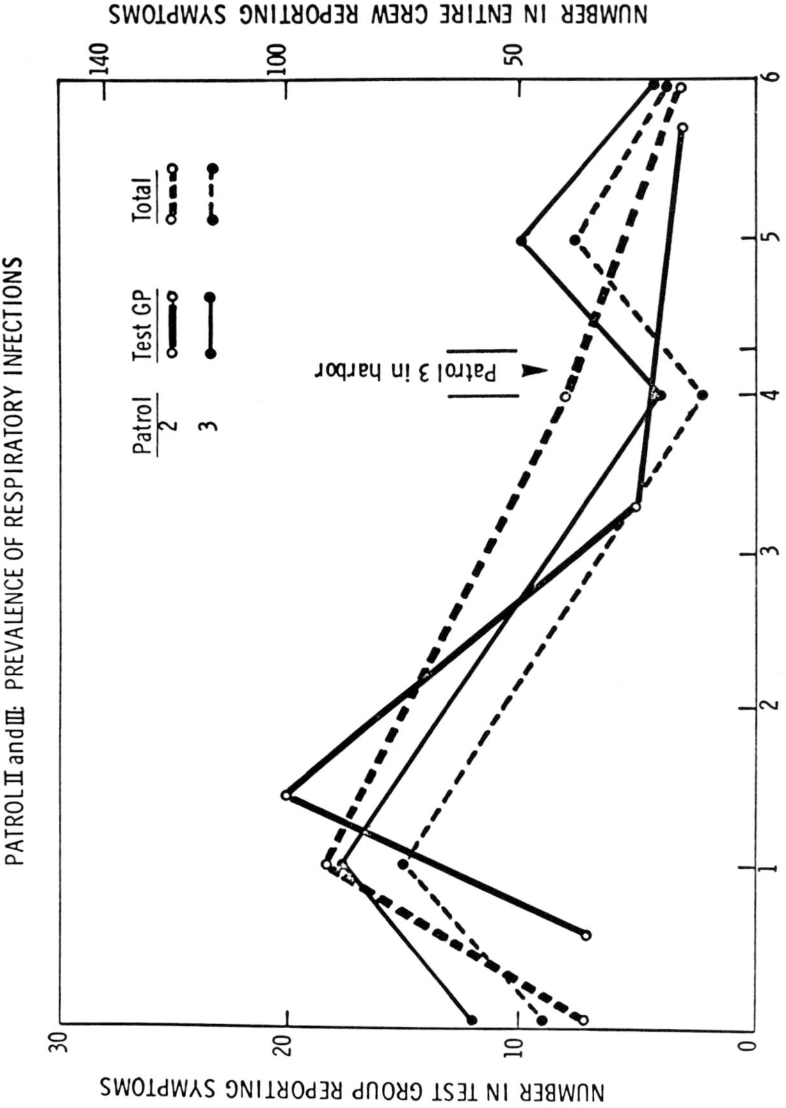
PATROL II and III: PREVALENCE OF RESPIRATORY INFECTIONS
Patrol
2
3
Test GP
Total
Patrol 3 in harbor
NUMBER IN ENTIRE CREW REPORTING SYMPTOMS
140
100
50
NUMBER IN TEST GROUP REPORTING SYMPTOMS
30
20
10
0
1
2
3
4
5
6
WEEKS AT SEA

Figure 8

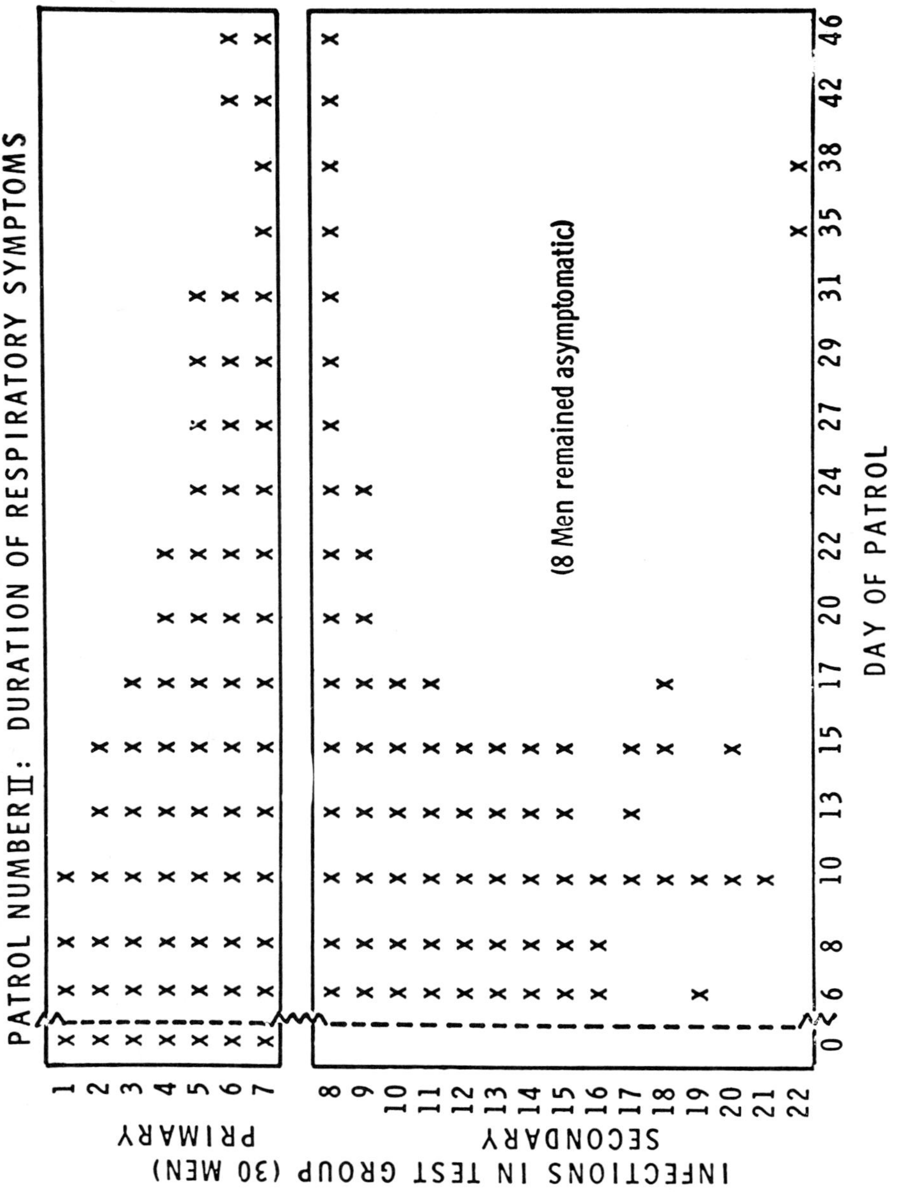
PATROL NUMBER II: DURATION OF RESPIRATORY SYMPTOMS
(8 Men remained asymptomatic)
DAY OF PATROL
0 6 8 10 13 15 17 20 22 24 27 29 31 35 38 42 46
INFECTIONS IN TEST GROUP (30 MEN)
PRIMARY
SECONDARY
1 2 3 4 5 6 7
8 9 10 11 12 13 14 15 16 17 18 19 20 21 22

Figure 9

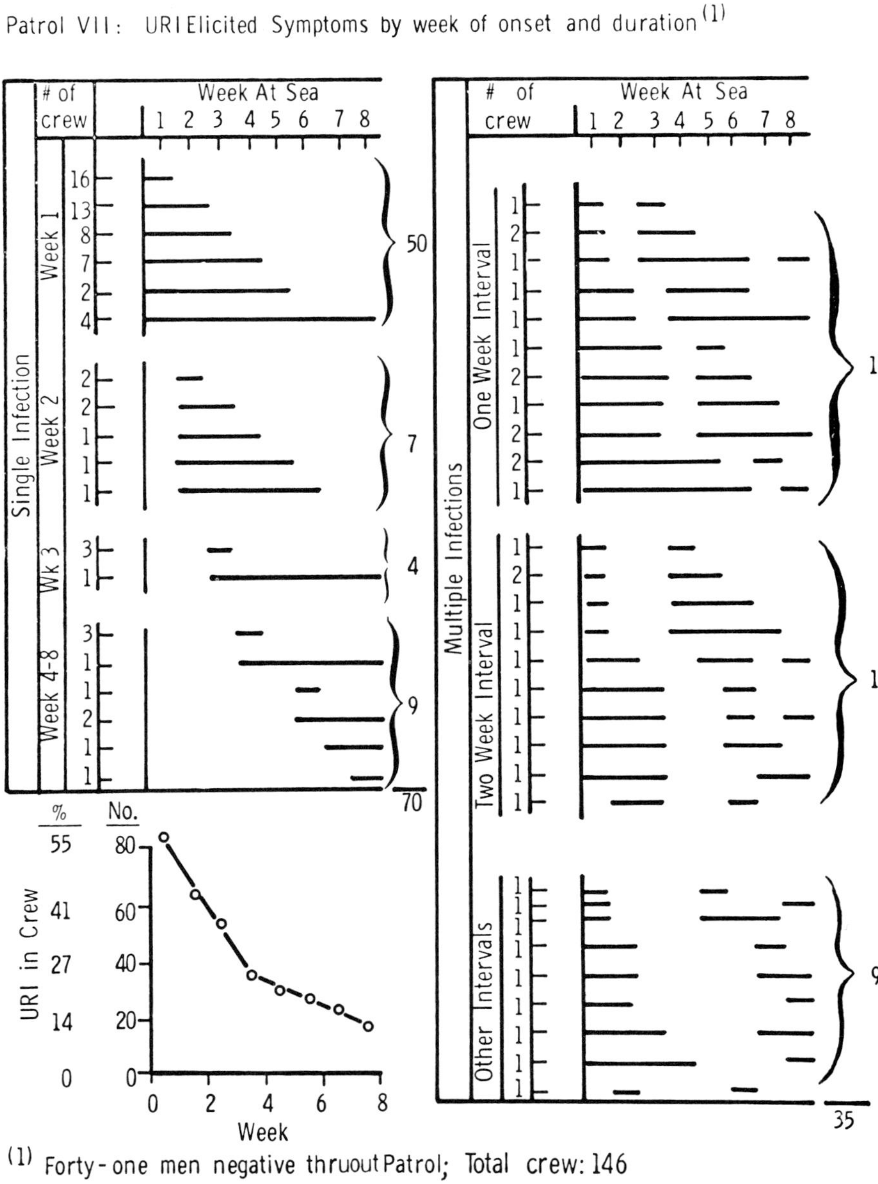

Patrol VII: URI Elicited Symptoms by week of onset and duration(1)
Single Infection
# of crew
Week At Sea
1 2 3 4 5 6 7 8
Week 1
16
13
8
7
2
4
50
Week 2
2
2
1
1
1
7
Wk 3
3
1
4
Week 4-8
3
1
1
2
1
1
9
70
Multiple Infections
# of crew
Week At Sea
1 2 3 4 5 6 7 8
One Week Interval
1
2
1
1
1
1
2
1
2
2
1
15
Two Week Interval
1
2
1
1
1
1
1
1
1
1
11
Other Intervals
1
1
1
1
1
1
1
1
1
9
35
%
55
41
27
14
0
URI in Crew
No.
80
60
40
20
0
0 2 4 6 8
Week
(1) Forty-one men negative thruout Patrol; Total crew: 146

Figure 10

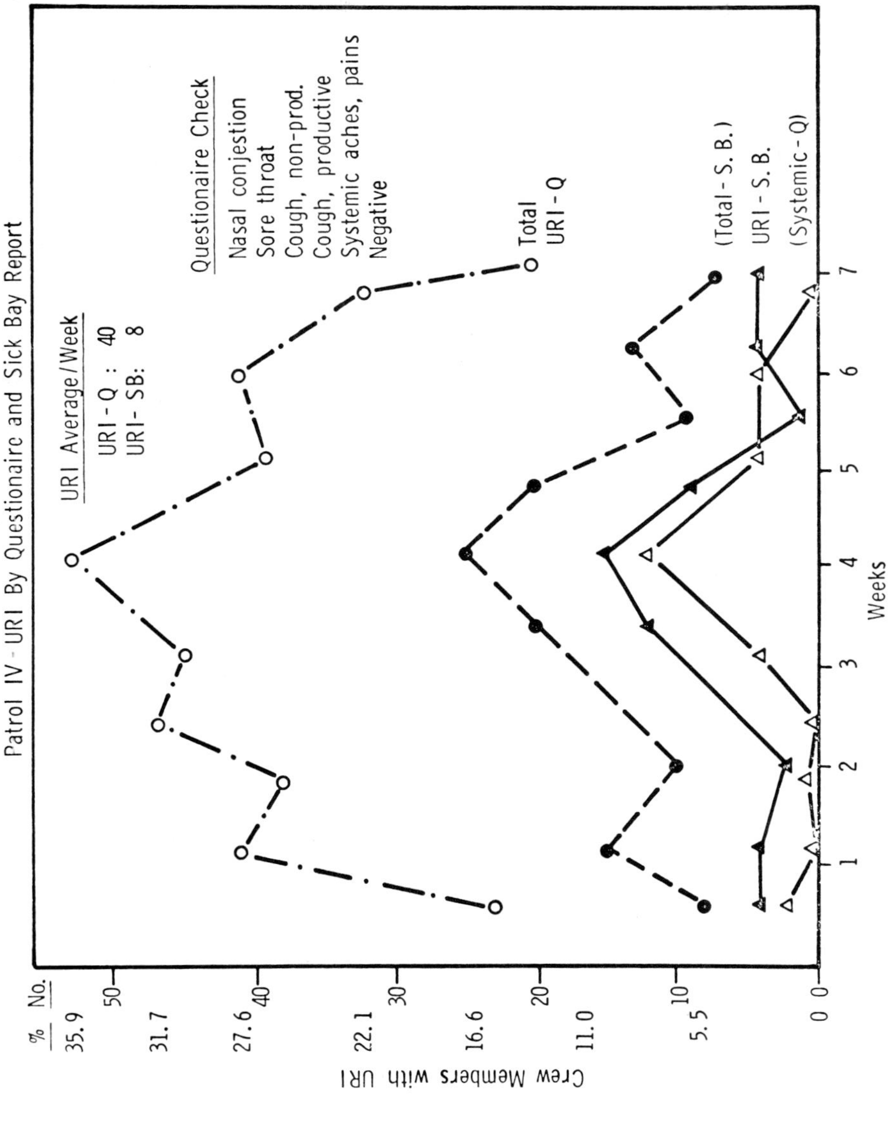
Patrol IV - URI By Questionaire and Sick Bay Report
URI Average/Week
URI - Q : 40
URI - SB: 8
Questionaire Check
Nasal conjestion
Sore throat
Cough, non-prod.
Cough, productive
Systemic aches, pains
Negative
Total URI - Q
(Total - S. B.)
URI - S. B.
(Systemic - Q)
Crew Members with URI
% No.
35.9
50
31.7
27.6 40
22.1 30
16.6
20
11.0
10
5.5
0 0
1
2
3
4
5
6
7
Weeks

Figure 11

Alteration of the pathogenic role of streptococci group C in mice conferred by previous exposure to ozone

David L. Coffin, V.M.D. and **Earl J. Blommer,** M.S.

National Air Pollution Administration, Cincinnati, Ohio 45237

ABSTRACT

Exposure of mice to ozone increases the possibility of their death from subsequent infection by aerosolized streptococci. Associated with the increased mortality is prolonged retention of intrapulmonary bacterial viability, a shortening of the lag phase and an increased rate of bacterial multiplication. The diminished lag phase and enhanced bacterial growth contribute to the slowing of the so-called bacterial clearance rate. Bacterial invasion of the blood appears to take place between the second and fourth days inclusively and equates with subsequent mortality. Bacteria present in the blood are in insufficient numbers to appreciably affect the total lung counts.

INTRODUCTION

It is now well established that a number of environmental influences alter the lung's ability to cope with artificially introduced bacteria (1). Manifestations of this phenomenon have been a prolongation of the so-called bacterial clearance or an increased mortality to infection from the introduced bacteria associated with exposure to noxious environmental condition. A model system employing enhancement of mortality as the determinative factor has been particularly successful in demonstrating the effects of oxidant air pollutants (2, 3, 4, 5).

It would seem that there is agreement among those who have worked with various environmental agents that the two phenomena, i.e. slowing of the clearance rate and enhancement of mortality, are related, such association having been previously noted for alcohol administration, and exposure to nitrogen dioxide and ozone (3, 4, 6). Bacterial clearance is generally thought to be a summation of those factors which physically remove bacterial cells from the lung and those which cause their loss of viability in situ. Although there is some difference of opinion as to the relative importance of these two factors in this phenomenon, recent work in mice, at least, indicates that the decline in bacterial viability plays the major role (7). Furthermore, this rapid loss of bacterial viability has been attributed to the activities of the pulmonary alveolar macrophage, presumably the first line of defence against introduced particulate (8).

Previous studies from this laboratory have indicated that there is a depression of the rate of bacterial clearance for mice associated with ozone exposure (3). Experiments conducted in rabbits subsequently examined by means of pulmonary lavage have shown a reduction in number and phagocytic activity of the pulmonary macrophage and a loss of potency in the lysosomal enzymes of these cells (9, 10). However, this association of slowed clearance and ozone appeared no longer so evident at lower concentrations of the gas which still yielded appreciable enhancement of mortality (11). While these observations might have been colored by lack of precision of our method imparted by variability of reaction of the ozone-treated animals, the difference noted in mortality as compared to clearance appeared so marked that such factors might be precluded.

A series of experiments were performed to extend the observations for bacterial decline or growth beyond the four-hour period to obtain data on the influence of ozone on clearance rate, bacterial growth and mortality.

MATERIALS AND METHODS

For these experiments groups of pathogen-free mice approximately 20 days of age were randomly divided into two sub-groups, one of which received a three-hour exposure to ozone (9), the other being reserved as ozone-free controls but residing in similar chambers. The ozone was generated by passing oxygen through a neon-tube silent-arc generator. The resulting effluent was mixed with CBR filtered room air to equal a total flow rate of 11·4 cubic feet per minute. This was conveyed into a stainless steel chamber with a volume of 11·4 cubic feet exclusive of funnels. Ozone was monitored by the neutral potassium iodide method. At the expiration of the three-hour exposure, the two groups were randomly intermixed in racks of individual exposure cells, placed in a similar chamber, and simultaneously exposed to a bacterial aerosol generated by a DeVilbiss* atomizer. The aerosol consisted of a washed 19-hour culture of group C streptococci diluted to conform to a turbidity scale previously standardized by cultural methods to achieve an initial inhaled dose of approximately $1{\cdot}5 \times 10^4$ per control mouse. After approximately seven minutes the fluid in the DeVilbiss was exhausted. Following this 15 minutes were permitted to elapse to allow for bacterial decline in the chamber. The animals were then removed and divided again into the ozone and control treatments. Each treatment group was then subdivided as follows: 30 to be reserved for observation of mortality to 12 days and 48 to be used for clearance and growth studies. The clearance sub-groups of

* Mention of company or product name does not constitute endorsement by the Department of Health, Education, and Welfare.

eight animals each were sacrificed immediately (0 time) and at four hours, one day, two days, three days and four days following the aerosol exposure. Cultures were performed on appropriate dilutions of whole lung homogenates, colonies counted and expressed in logs as numbers of viable bacteria (cultivable units) per total lung. Blood cultures were performed during necropsy at certain ozone levels in an ancillary study to be reported in detail elsewhere.

RESULTS

Following the introduction of pathogenic streptococci into normal (non-ozone exposed) mice, there was a precipitous decline in the number of cultivable organisms most marked at four hours. Median residual values at this point amounted to approximately 3 per cent of those at 0 time. Subsequently there was slight reduction amounting to approximately 2 per cent to the one-day period after which values tended to plateau or slightly elevate.

Administration of ozone at sufficient concentration markedly altered the number of organisms cultivable at four hours. Exposure to concentrations of 1 p.p.m. or above for three hours resulted in an actual gain in bacterial numbers, the highest ozone concentration used (5 p.p.m.), this gain amounted to nearly four times the 0 time level. At levels of 0·22 p.p.m. or below, however, bacterial counts for the ozone treatments could not be distinguished from controls. Similar results have been noted in an earlier study reported elsewhere (11). The bacterial loss or gain at four hours is summarized in Figure 1.

For periods beyond four hours marked variability in the numbers of cultivable cells was noted in all groups save the higher ozone concentrations yielding mortalities approaching 100 per cent. The counts ranged from 10^1 (the lowest concentration counted) to 10^{16} or above in both the controls and the ozone treatments at the four-day period. For the ozone treatment groups yielding mortality approaching 100 per cent, however, there was very little variability and counts ranged within the span of a log or two even at four days (Figure 2). Despite this variability when median values are plotted, a dose related association of the number of cultivable bacteria and ozone concentration is evident. Furthermore, mortality appears directly related to each. Regression analysis of the data indicated a significance of $P<0{\cdot}001$ and $P<0{\cdot}01$ respectively for the relationship of bacterial counts and mortality to ozone concentrations. The data for these experiments are summarized in Figure 3.

Bacterial invasion of the blood stream begins on the second day and appears largely completed by the fourth day when positivity for blood cultures roughly equates with mortality for both controls and ozone-treated animals.

DISCUSSION

From results of this experiment it is clear that ozone exposure in mice influences the rate of bacterial decline or growth in the lung, invasion of the blood and subsequent mortality. In this series of experiments as in a preceding one, while the four-hour bacterial clearance rate was markedly altered at high levels of gas exposure, there was little or no reproducible difference from control at lower levels of exposure normally associated with enhanced mortality. Furthermore, the slowing of the clearance rate at the higher levels of exposure patently contributed to a rapid growth phase of the organism, since an actual increment of the number of organisms occurred at ozone concentration above 1 p.p.m. It therefore seems reasonable to assume that slowing below this level was also made up of a composite of bacterial decline in diminishing amounts, and bacterial growth at increased rates. Loss of cultivable bacteria from the lung is obviously accomplished by those factors which physically remove the bacterial cells such as mucus flow and those which cause a loss of viability in situ such as passive decay and active destruction by defensive mechanisms. In ozone exposure defensive mechanisms appear to be depressed sufficiently to permit slowing of bacterial kill and an onset of growth the sum of which is reflected in the slowed rate observed at the four-hour period. It should be noted that other studies from this laboratory have shown that ozone exposure produced a meaningful but transient loss of competence by pulmonary aveolar macrophages obtained by pulmonary lavage. These alterations have consisted of a reduction in number and phagocytic activity (3, 9). There has also been a reduction in activity of the lysozymal enzymes (phosphatase, beta fluouronidase, and lysozyme) (11). When studied by electron microscopy, ozone exposure at 5 p.p.m. was observed to have resulted in nonspecific but profound morphologic alterations, with dilatation of the endoplasmic reticulum and perinuclear space, swelling of mitochondria, formation of myelin figures, autophagic vacuoles, and ultimately cell lysis (12). Furthermore, reduction of the activity of lysozyme not associated with cells has also been noted (13). It is likely that these phenomena are largely instrumental for the slowed rate of bacterial kill and growth acceleration noted in the studies reported here.

Beyond the four-hour period acceleration of bacterial multiplication in the lung might be explained by a number of factors. For instance one factor to be considered is that the larger four-hour residual dose simply provides a larger base from which to commence the growth phase. While it does not appear possible to explain the entire bacterial multiplication difference on this basis, it is entirely possible that the increased base resulting from reduced clearance coupled with a gradually declining continuum of the macrophage incompetence conferred by ozone might largely account for the sudden growth in the treated animals.

As can be seen from the data, while there is a general correlation between bacterial lung counts, mortality, and ozone concentration, bacterial numbers continue to mount where mortality obviously cannot (100 per cent). Such a phenomenon can possibly be explained in these experiments by the fact that bacteria are growing wildly in the lungs of animals already predestined for death because of co-existing septicemia which also appears meaningfully related to ozone exposure. Preliminary review of tissue sections in a study to be reported elsewhere indicates that the pneumonia produced in this model is generally focal in nature and is frequently accompanied by intense bacterial accumulation in alveolar spaces not directly associated with inflammatory cells. One might speculate that such growth may be fostered by the presence of odema fluid induced by ozone at levels above 1 p.p.m. However, the enhanced intrapulmonary bacterial growth, invasion of the blood, and mortality seen at non-edemagenic concentration of ozone would appear to be related to the action of the gas on a specific defensive mechanism. Evidence from this laboratory suggests that impairment of the macrophage system may be the key factor in such phenomena.

REFERENCES

1. Kass, E. H., Green, G. M. and Goldstein, E. (1966). Mechanisms of anti-bacterial action in the respiratory system, *Bact. Rev.*, **30,** 488–497.
2. Miller, S. and Ehrlich, R. (1958). Susceptibility to respiratory infection of animals exposed to Ozone I. Susceptibility to *Klebsiella pneumoniae, J. Infect. Diseases,* **103,** 145.
3. Coffin, D. L., Blommer, E. J., Gardner, D. E. and Holzman, R. (1967). Effect of air pollution on alteration of susceptibility to pulmonary infection, *Proceedings of the 3rd Annual Conference on Atmospheric Contamination in Confined Spaces*, Dayton, Ohio, May 1967.
4. Ehrlich, R. (1966). Effect of nitrogen dioxide on resistance to respiratory infection, *Bact. Rev.*, **30,** 604.
5. Coffin, D. L. and Blommer, E. J., (1967). Acute toxicity of irradiated auto exhaust indicated by mortality from streptococcal pneumonia, *Arch. Environ. Health*, **15,** 36.
6. Stillman, E. G. (1923). The presence of bacteria in the lungs of mice following inhalation, *J. Exp. Med.*, **38,** 117–126.
7. Green, G. M. and Kass, E. H. (1964). Factors influencing the clearance of bacteria on the lungs, *J. Clin. Invest.*, **43,** 769.
8. Green, G. M. and Kass., E. H. (1964). The role of the alveolar macrophage in the clearance of bacteria from the lungs, *J. Exp. Med.*, **119,** 167.
9. Coffin, D. L., Gardner, D. E. and Holzman, R. S. (1968). Influence of ozone on pulmonary cells, *Arch. Environ. Health*, **16,** 633–636.
10. Hurst, D. J., Gardner, D. E. and Coffin, D. L. *Effect of Ozone on Lysozymes*, in preparation.
11. Coffin, D. L., Blommer, E. J. and Wolock, F. J. (1968). Alteration of bacterial lung clearance rates by air pollutants, presented at the 9th Air Pollution Medical Research Conference, Denver, Colorado, July 1968.

12. Huber, G. L., Gardner, D. E., and Coffin, D. L. *Electron Microscopy Alterations in Rabbit Macrophages Following Experimental* in vivo *Exposure to Ozone*, in preparation.
13. Holzman, R. S., Gardner, D. E. and Coffin, D. L. (1968). *In vivo* inactivation of lysozyme by ozone, *J. Bacteriol.*, **96**(5), 1562–1566, 1968.

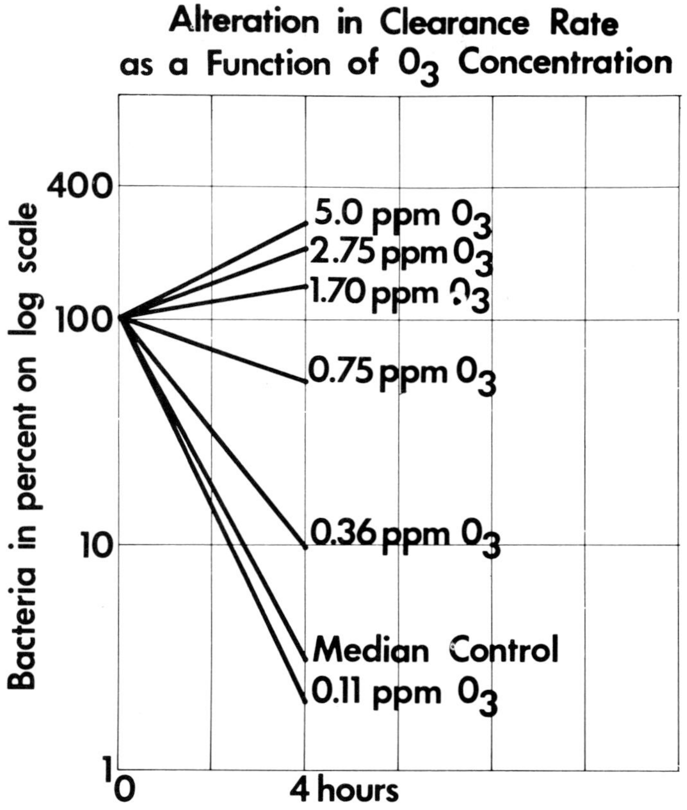

Figure 1

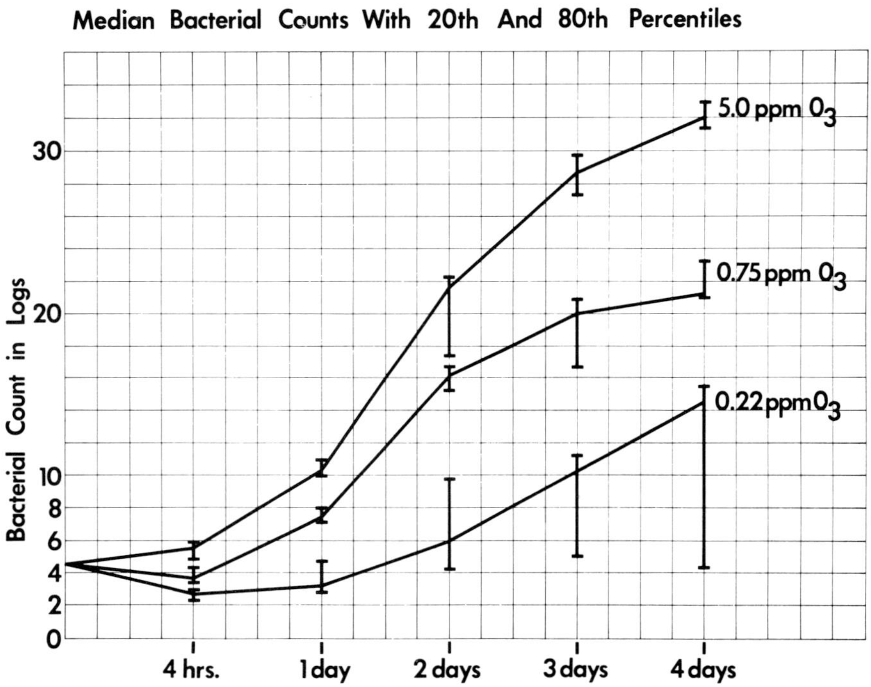
Median Bacterial Counts With 20th And 80th Percentiles
Bacterial Count in Logs
30
20
10
8
6
4
2
0
5.0 ppm O_3
0.75 ppm O_3
0.22 ppm O_3
4 hrs.
1 day
2 days
3 days
4 days

Figure 2

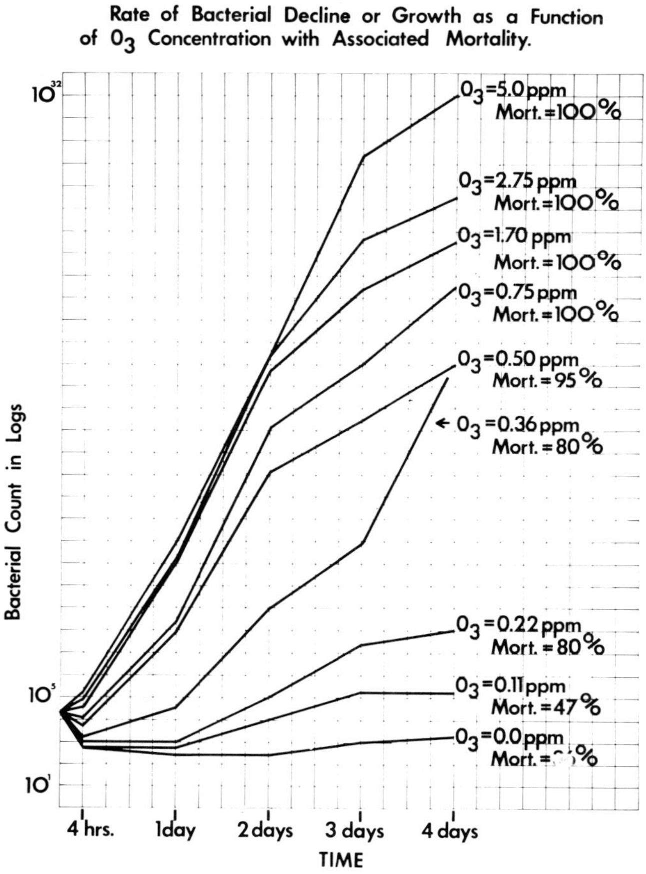
Rate of Bacterial Decline or Growth as a Function
of O_3 Concentration with Associated Mortality.
10^{32}
10^{5}
10^{1}
Bacterial Count in Logs
O_3=5.0 ppm
Mort.=100%
O_3=2.75 ppm
Mort.=100%
O_3=1.70 ppm
Mort.=100%
O_3=0.75 ppm
Mort.=100%
O_3=0.50 ppm
Mort.=95%
O_3=0.36 ppm
Mort.=80%
O_3=0.22 ppm
Mort.=80%
O_3=0.11 ppm
Mort.=47%
O_3=0.0 ppm
4 hrs.
1day
2days
3 days
4 days
TIME

Figure 3

Airborne pathogenic bacteria in a tissue transplant unit

J. C. Gould

Central Microbiological Laboratories, Edinburgh

SUMMARY

The unit to be described has been designed to maintain as near as possible a pathogen-free environment within a 'clean area' for the treatment of patients requiring transplantation or whose circumstances greatly increase their susceptibility to infection.

The ventilation system effectively excludes bacteria from the outside environment and each individual patient can be isolated in a cubicle from the moment of entry into the unit until his discharge. Rigorous procedures attempt to exclude the introduction of bacteria on fomites, food and other articles which are continually being brought into the clean area; both staff and patients are examined for the presence of known pathogens on their surfaces. Regular monitoring of both personnel and the internal environment of this unit is carried out to detect pathogenic or potentially pathogenic bacteria and when found these are traced.

Results are now available following 15 months' continuous operation of the unit. Low bacterial counts can be maintained within the central clean area, the theatre and the patient's rooms under normal conditions of activity, but these rise when there is increased activity of personnel. The bacteria isolated from the airborne dust and surfaces are derived from the clothes of the staff and their body surfaces, and from the bedclothes and surfaces of the patients; these differ. In only a few instances have organisms of potential pathogenic significance been isolated from other sources. usually due to defects in the procedures for bringing items into the unit.

The bacteria derived from the patients have been strictly retained within the individual's room and there has been virtually no evidence of spread from one room to another nor to other parts of the clean area of the unit.

These results are related to the incidence and nature of infections that have occurred among the patients treated in the unit.

INTRODUCTION

The number of patients unable to resist infection as a consequence of their treatment is increasing: in particular this includes patients for transplantation surgery and others who are on immuno-suppressive drugs for auto-immune disease and neoplasms. It is probable that the treatment of such patients in open hospital wards exposes them to an unnecessary risk of infection and they must be isolated for their protection. It is the particular function of the Tissue Transplantation Unit (Woodruff *et al.* 1968) to provide this protection and to this end the Unit is designed primarily to minimize the risk of airborne infection and involves strict isolation of the patient so that barrier nursing supplemented by antiseptic and aseptic measures can be fully effective. It is also hoped that this Unit will enable the assessment of the practical importance of a number and variety of anti-infection procedures, either alone or in combination.

The sources of organisms pathogenic to man are most frequently human: thus infections most frequently derive from other persons or from the patient himself. Under ordinary conditions other groups of bacteria, such as saprophytes, rarely if ever produce infections, but with patients at particular risk these organisms can become opportunists and assume pathogenic significance; such patients are also more easily invaded by commensals and pathogens.

The routes of spread of microflora from these sources may be direct; the patient may infect himself, or there may be direct transfer from his attendants. Otherwise spread may be indirect through reservoirs which include intermediate persons, dust, air, droplets and droplet nuclei, water, food, fomites, and living vectors such as insects. The protection of the patient from contamination and thus potential infection must take into account methods capable of control of spread of micro-organisms by these routes.

The Tissue Transplantation Unit

This comprises six isolation suites and a pair of operating theatres which come off a central clean area (Figure 1). Access to this central area is on three sides by open corridors enabling supply and return as well as facilities for various types of examination of the patient, television viewing and visitor facilities.

The air ventilation system pumps fresh, humidified, warmed or unheated air through primary and secondary filters into the Unit. There is no extract in the central area. The secondary banks of filters are 99·9 per cent proof against particles down to 0·5 micron. The theatre air supply is delivered at 2400 cubic feet per minute and is equivalent to 25 air changes per hour: the clean area supply is at 3,320 cubic feet per minute and the patient area at 2300 cubic feet per minute equivalent to six changes per hour. This system maintains highest pressure (P4) in the central clean area and this pressure progressively reduces to the theatre (P3), and patients' rooms (P3), to the changing accommodation and patients' bathrooms (P2) and the outside corridors (P1) which are equipped with extraction. The pressure is lowest in the soil return corridor situated in the basement so that there is a minimal danger of reflux or organisms into the patients' room during the use of the hoists for soil return.

Siting of the services allows maintenance without entry into the central clean area.

Procedures

The patient enters through the bathroom of his isolation suite, bathes and applies antiseptic cream (Hexachlorophane). He is then isolated in his room clothed in sterilized garments. The bathroom is now decontaminated with an antiseptic-detergent solution, and after time has been allowed for the air ventilation scheme to remove organisms from the bathroom, the doors to the outside corridor are finally locked.

Staff enter the clean area via their changing accommodation where they decontaminate by washing, discard their outdoor garments and put on clean sterilized clothing.

Soil return and all other materials from the isolation suites are discharged via a hoist to the basement. The delivery of this hoist allows very slow air movement which at all times is overcome by the down draught from high to low pressure.

Food is supplied from the main hospital kitchens in the containers in which it has been cooked. The outside surfaces of these and all bottles containing fluids required in the central area are washed with clean cloths soaked in a phenolic detergent solution prior to being passed through a double hatch. The food is served on sterile crockery supplied from the sterile supply. All other articles for use in the central clean area are supplied to the clean supply store in triple wrapping. The outer dust cover is removed as the packings are transferred through the clean supply hatches.

There is complete disinfection of each suite after the patient leaves.

BACTERIOLOGICAL STUDIES

Measurement of the microbial content of the environment of the Unit serves as a check on the efficiency of the air ventilation system and as an early warning system to detect bacterial contamination resulting from faults in procedure. The following monitoring studies have been regularly carried out (Table 1):

1. Estimation of the total bacterial count and occurrence of specific pathogens in both airborne and settling dust.
2. The examination of fomites and surfaces such as walls for specific pathogens.
3. An examination of antiseptic and washing solutions used within the Unit for the presence of micro-organisms.
4. The noses, throats, skin flexures, hands, faeces and urine of members of the staff have been examined for potential pathogens at regular intervals. Particular attention has been paid to the occurrence of antibiotic-resistant Gram-negative bacilli in the faeces of these persons.

Before admission all patients have been examined in a similar manner and thereafter as indicated during their stay in the Unit. Urine and exudates were examined daily.

Methods

Airborne dust was sampled by means of a modified Du Boy and Crisp sieve sampler taking air at between 25 and 30 cubic feet (M^3) per minute through 400 holes onto the surface of agar plates. Bacteria-carrying particles were trapped on the surface and recovery was approximately 60 per cent for particles within the range 1–20 microns.

Settling dust was sampled by exposing agar plates for known periods of time at different sites. This was a more convenient method for routine monitoring in the Unit since the machines were noisy and prolonged sampling became irksome to the patients and staff.

Air sampling by means of a slit sampler was carried out in the patients' rooms by using adaptors attached to vents in the partition separating these rooms from the outside corridor.

The culture plates from the air-sampling experiments were incubated after sampling and colony counts were made (total counts), each colony being regarded as corresponding to a 'bacteria-carrying particle'. Exposures were adjusted so that no more than 200 colonies per plate were present to avoid gross counting errors.

Plates cast with blood agar were generally used to obtain total bacterial counts. Selective media were used for the isolation of specific organisms, e.g. salt milk agar for *Staphylococcus pyogenes*, and Dettol agar for Pseudomonas.

Fomites were sampled using a variety of techniques, e.g. dry and wet swabs; velvet impression pads; agar sausages and adhesive tape.

Swabs and impression preparations from members of staff and patients were cultured on blood agar or other medium as indicated and all plates were incubated in air, in carbon dioxide and anaerobically.

Further bacteriological examinations

Colonies of Staphylococci were tested for coagulase production by both the slide test (Cadness-Graves, 1943) and the tube technique of Fisk (1940). Coagulase-positive strains were subcultured in nutrient broth and tested for their susceptibility to bacteriophage and antibiotics. Coagulase-negative strains were tested for their biochemical activities and classified according to Cowan and Steel (1965).

Colonies of Gram-negative coliform bacilli were subcultured and identified by the appropriate biochemical tests. Subcultures of Pseudomonas were typed using a set of bacteriophages (Gould and McLeod, 1960).

Results

The results of monitoring the 78 members of staff are shown in Table 2. *Staph. pyogenes* was by far the most frequently isolated pathogen from the external surfaces of these persons.

No antibiotic-resistant *Esch. coli* or other coliform bacillus was isolated from the samples of faeces examined and in only two instances were strains of Escherichia isolated from the urine. *Salmonella typhi-murium* was isolated from the faeces of one nursing auxiliary. Respiratory-tract pathogens were isolated on a number of occasions, usually from asymptomatic persons. Occasionally similar organisms were recovered amongst the airborn microflora of the environment of the Unit.

The bacteria recovered from the dust of the air in the Tissue Transplant Unit were predominantly cocci—Micrococcus, *Staphylococcus albus, Staphylococcus epidermidis,* Sarcina and Aerococcus. Corynebacteria were also fairly numerous (Table 3). These organisms were probably derived in the main from the skin and clothes of the medical attendants and patients. Larger numbers of *Staph. epidermidis* and *Staph. pyogenes* were isolated in the patients' rooms than in the central clean area or theatre and were most probably derived from the patients and their bed clothes.

The number of airborne and settling bacteria-carrying particles was considerably less than in the general hospital environment and differs in the almost complete absence of fungal and bacillary spores, which indicates an effective barrier by the filtration mechanism (Table 4).

The number of airborne and settling particles carrying recognized pathogens was very small, and on the greater proportion of occasions on which examination was carried out, no such pathogens were isolated in the clean area and even in the patients' rooms isolation was rare and intermittent.

The probable source of these environmental pathogens and opportunists is given in Table 5. About half of the strains of *Staph. pyogenes* isolated corresponded to strains known to be carried by members of staff. These carriers were regularly treated with suppressive antiseptic preparations, but in spite of this the organism could be recovered by swabbing of the nares and flexures in many individuals. In view of this special measures were taken to study the dissemination of organisms by staphylococcal carriers.

Measurement of dissemination by staphylococcal carriers

Since approximately 30 per cent of the population are persistent carriers of *Staph. pyogenes* (Gould and McKillop, 1954) it was to be expected that a corresponding proportion of the medical and nursing staff employed in this Unit would be carriers. It is difficult if not impossible to completely suppress the organism in persistent carriers by means of antiseptic sprays and creams applied to the nares and skin, although this may be helpful in controlling dissemination of the organism, and it is quite clearly out of the question to exclude them from working in the Unit since carriage is long-term. Not all such carriers, however, are a hazard to the community since only a proportion are disseminators, (Eichenwald *et al.* 1960). Persistent carriers on the nursing staff of this Unit were therefore measured for dissemination of their staphylococci.

This was done by using an air sampling box (Gould, 1959) in which the individual carriers-out simulated movements for a given time and from which the air is sampled into a Casella slit sampler. In this way the number of particles carrying *Staph. pyogenes* which have been disseminated may be counted and their persistence in the environment measured. These observations were supplemented by measurements of the number of organisms obtained from the surfaces of the individual by touching dry and wet surfaces and by washing the hands. The results of these measurements in a selection of known and untreated nasal carriers are compared with measurements on three carriers among the whole-time nursing staff of the Unit and the results shown in Table 6.

The number of organisms present on the hands of these nurses was small and this must be attributed to the frequent washing in residual antiseptics such as Hexachlorophene; this in addition to topical antiseptics (Soframycin spray and Chlorhexidine nasal cream) applied to the surface available for dissemination. Since this is the case it seems a safe procedure to allow persistent nasal and perineal carriers who are not disseminators to work in the Transplant Unit and the number of *Staph. pyogenes* which have been isolated during the monitoring period supports this policy.

Results of monitoring patients

A high proportion of the patients for transplantation were found to be heavily colonized with *Staph. pyogenes* before admission to the Unit (Table 7): this in part may be the result of their long exposure to hospital procedures prior to transplantation. Some of the strains isolated were highly resistant to antibiotics and were of different phage type to the strains known to colonize members of the staff.

Staphylococcal infection is common among patients for transplantation and this has been confirmed in the present series. Several of the patients in this series had had previous staphylococcal infection, e.g. infection of their arterio-venous shunts, and boils and other superficial infections. Of the nine wound infections in this series, seven were due to *Staph. pyogenes* and in six of these the organism was the same as that colonizing the nose and other areas of the patient. This indicates that the patient was the source of his own infection and at no time was *Staph. pyogenes* other than the patient's own carrier strain, isolated from the rooms of the patient. The remaining two infections, due to *Staph. albus* and *Esch. coli* respectively, were also most probably due to an autogenous source.

There is thus no evidence that airborne organisms were the source of any of the patient's infections. Two patients had post-operative respiratory infection from which *Haemophilus influenzae* was isolated. These strains were non-capsulated and small numbers were isolated from the air and immediate environment of these patients during the acute phase of the illness. One member of staff had a respiratory tract infection from which *H. influenzae* was isolated, but it is problematical whether this had any connection with either of the two patients' infections. There were also a few minor superficial infections among the patients: two cases of *Otitis externa* due to

Streptococcus, a skin rash from which no organism was isolated, and two with staphylococcal boils. It is most unlikely that the source of these infections was airborne. There is no evidence of infection of any kind with opportunist organisms. Pseudomonas and Achromobacter were detected on several occasions in the clean area and once in a patient's room on the surface of the bedside locker. The source in at least two of these instances was the ice from the kitchen refrigerator which had been added to the sterile drinking water provided for the patient. This was a breach in procedure, which indicates the potential hazard when the strict rules laid down are not followed and reinforces the need for constant supervision.

Twelve of the 14 patients had organisms in their urine after operation; in most cases these organisms were *Esch. coli*, *Proteus* species and occasionally *Streptococcus faecalis*; these organisms were most frequently fully sensitive to antibiotics and it seems certain that these were derived from the patients' own surfaces. In one patient there was an associated *Esch. coli* infection of his wound, but none of these organisms were ever isolated from the air of the Unit.

REFERENCES

Cadness-Graves, B., Williams, R., Harper, G. J. and Miles, A. A. (1943). *Lancet*, **i,** 736.

Cowan, S. T. and Steel, K. J. (1965). *Manual for the Identification of Medical Bacteria.* Cambridge University Press.

Eichenwald, H. F., Kotsevalov, O. and Fasso, L. A. (1960). *Amer. J. Dis. Child.*, **100,** 161.

Fisk, A. (1940). *Brit. J. Exp. Path.*, **21,** 311.

Gould, J. C. (1959). Thesis for M.D. University of Edinburgh.

Gould, J. C. and McKillop, E. J. (1954). *J. Hyg. Camb.*, **52,** 304.

Gould, J. C. and McLeod, J. W. (1960). *J. Path. Bact.*, **79,** 295.

Woodruff, M. F. A., Nolan, B., Bowie, J. H. and Gould, J. C. (1968). *Lancet*, **i,** 905.

TABLE 1

Bacteriological monitoring procedures

	Patients		*Medical, nursing and auxiliary staff*	*Domestic and maintenance*	*Internal environment*
Before entry	Surfaces Urine Faeces Lesions	*Weekly*	Surfaces Urine Faeces Hands	Periodically as required	Airborne dust Settling dust
		Exclusion	If lesions occur		Surfaces and fomites
On entry					
	Decontamination	Discard outdoor clothing, wash and put on sterile clothing. Use antiseptic precautions on surfaces.			Plumbing Wash and cleansing materials Dialysis equipment Humidifiers and filters.

TABLE 2

Bacteria isolated by monitoring of medical staff and auxiliaries

Number of staff examined	*Number yielding*				
	Staphylococcus pyogenes		*Gram-negative bacilli*	*Pseudomonas*	*Other potential pathogens*
78	*Frequently*	*Rarely*			
Nose	20	4	1	0	Haemophilus 1
Throat	1	0			Haemophilus 1 *Streptococcus pyogenes* 4 Pneumococci 1
Perineum	2	0	–		
Hands	0	3	4		
Urine	–		2		Streptococcus 1
Faeces	–				*Salmonella typhi-murium*
Lesions Boils	2				
Ear	Anaerobic Streptococcus				

TABLE 3

Frequency of isolation of various bacteria from the airborne and settling dust

	Proportion of total counts in	
	Patients' rooms %	*Clean area* %
Micrococcus	45	70
Staphylococcus albus	25	10
Corynebacterium	10	5
Sarcina	3	10
Bacillus, Clostridium	1	< 0·1
Aeromonas, Achromobacter, Neisseria, Streptococcus, *Staphylococcus pyogenes*, Haemophilus, Pseudomonas, Proteus	< 0·1	< 0·1

TABLE 4

Measurements of airborne and other environmental bacteria

	Clean area	*Empty patients' rooms*	*Occupied patients' rooms*	*Theatre*	*Open corridors*	*Comparable areas in general hospital*
Airborne dust						
Bacteria-carrying particles/M^3	8–140	12–72	30–210	3–55	265–1700	110–835
Settling dust						
Bacteria-carrying particles per M^2/hr.	80–320	24–80	80–800	15–50	2880	290–4800
Fomites	Organisms less frequently isolated and in smaller numbers than in general hospital.					
Antiseptic washing solutions	Pseudomonas, Achromobacter and Coliform bacilli isolated on occasion.					

TABLE 5

Isolation of pathogens and opportunists and probable sources

Organism	*Number of occasions isolated*	*Probable source*
Staph. pyogenes	14	7/14 Medical attendants
	Airborne and settling dust	4/14 Patients
	Fomites	
Ps. pyocyanea	8	Refrigerator ice
	Settling dust	Antiseptic wash-water for floors
	Fomites	
Achromobacter sp.	2	Water and antiseptic solutions
	Settling dust	
S. typhi-murium	1	Contact of known case
Proteus sp.	5	Infected urine of patients
Esch. coli	6	Infected urine of patients
	Settling dust	
	Fomites	
Haemophilus influenzae	2	Respiratory secretions of patient and staff
	Airborne dust	
Pneumococcus	1	Respiratory secretions of patient and staff
Streptococcus pyogenes	1	

TABLE 6

Measurement of dissemination of Staphylococcus pyogenes from carriers

	Number of Staph. pyogenes isolated (viable counts)	
	Untreated carriers	*TTU staff carriers*
Nose swab	+++	+++
Dissemination by contact		
From washing of hands	10,000–790,000	0–200
Number of organisms transferred to sterile beads	0–950	0
Number of organisms transferred onto surface of agar plates	3–400	0–6
Body movement in cabinet for 5 minutes in outdoor clothes	26–570	0–15

TABLE 7

Bacteriology of patients' infections

Patient	Urinary tract	Wound infection	Other infection	Staphylococcal carriage in nose and flexures
1	*Esch. coli* 10^5 Proteus 10^5	*Staph. pyogenes* AR 47/53/54/77	None	*Staph. pyogenes*
2	*Staph. albus* 10^4	None	None	—
3	Proteus 10^5	*Staph. albus*	None	*Staph. pyogenes* 3a/3c/55/71 AS acquired AR 47/5 temporarily
4	*Esch. coli* 10^6 AR SR TR NeoR KS CS	None	None	—
5	None	*Staph. pyogenes* AS 55 acquired	Aural infection *S. viridans* *St. albus*	*Staph. pyogenes*
6	*Esch. coli* 10^5 *Str. faecalis* 10^6	None	Oral *Moniliasis* Coliform bacilli in dialysing fluid *Staph. pyogenes* in A.V. shunt AS 52/52a/80	
7	*Esch. coli* 10^5 Proteus 10^4	None	None	—
8	*Esch. coli* 10^{5-8} AR SS TR KS CS CephR CycloR	None	None	—
9	None	*Staph. pyogenes* AS 53/54/75/79	Nasal boil *Staph. pyogenes* Conjunctivitis Resp. infection	*Staph. pyogenes*
10	*Esch. coli* 10^5	*Esch. coli* *Str. faecalis*	None	*Staph. pyogenes* 3a/AS acquired AR NT
11	Coliform bacilli 10^7 *Str. faecalis* 10^5	*Staph. pyogenes* AS 6/47/54/77/85a	Old A.V. shunt *Staph. Pyogenes*	
12	Proteus 10^6	*Staph. pyogenes* AS 6/47/54/77/85a	Aural infection γ-streptococcus Resp. infection *Haem. influenzae* Conjunctivitis *Staph. pyogenes*	*Staph. pyogenes*
13	None	*Staph. pyogenes* AR 53/54/77/ 52/52a/80	Infected A.V. shunt *Staph. pyogenes*	*Staph. pyogenes*
14	*Esch. coli* 10^8	*Staph. pyogenes* AR 52/54/81	None	*Staph. pyogenes*

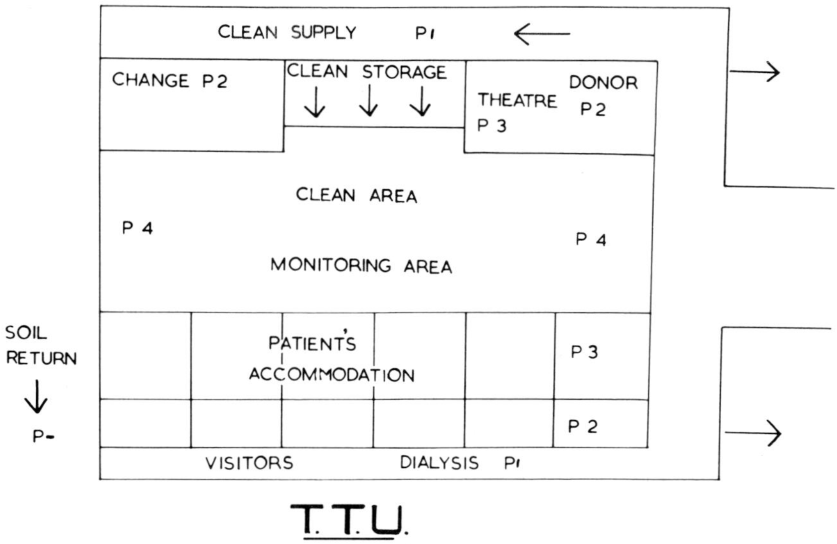
CLEAN SUPPLY P1
CHANGE P2
CLEAN STORAGE
THEATRE
P 3
DONOR
P 2
CLEAN AREA
P 4
P 4
MONITORING AREA
SOIL
RETURN
P-
PATIENT'S
ACCOMMODATION
P 3
P 2
VISITORS
DIALYSIS P1
T.T.U.

Figure 1

A look at dental aerosols

W. J. Hausler

University of Iowa, Iowa 52240

In the last ten years patterns of aerosols containing micro-organisms around a patient undergoing an operative dental procedure have been investigated, established and reported. The main impetus for these investigations was the development and practitioner use of newer operative instrumentation such as the air-turbine handpiece.

The air-turbine handpiece has been accepted by the dental profession as an efficient method of cutting tooth structure. It has provided an additional element of patient· comfort in its rapidity of movement as well as allowing more efficient utilization of professional time by reducing treatment time and allowing an increase in the number of patients treated. The use of this handpiece, which has operation speeds in excess of 100,000 r.p.m., has become as common as the dental mirror.

Early investigations were concerned with hearing loss of the practitioner due to his continued proximity to the characteristic high noise level of the turbine (Bernier and Knapp) (1) and also the effects of inhalation of aerosolized turbine lubricating oil (Bernier, Knapp and Boyers) (2). Although no definitive epidemiological investigations to our knowledge have been undertaken, the literature does reveal that a definite hearing range reduction has been known to occur because of their use of this handpiece (1, 2). The same can be said for infectious diseases contracted by dentists through treatment of patients suffering from various diagnosed or undiagnosed diseases.

Studies of aerosol clouds have revealed that an infectious disease problem has the capability of existing, but to our knowledge no definitive epidemiologic studies have been conducted involving the patient, handpiece or other aerosol-producing instrument, dentist and other patients. Dentists report that they have had more upper respiratory infections and more skin infections of the face since using this type of equipment, but so far no quantitative data have been reported to substantiate this claim. In 1910 Flick (3) wrote an article in regard to the management and treatment of tuberculosis patients. He stated that 'the air which the tuberculosis consumptive exhales is not dangerous except when it carried off broken-down tissue, or in forced breathing'. The dentist is involved not only with the tuberculosis-type patient but also those patients infected with agents commonly spread by the droplet route. The undiagnosed cases of contagious diseases are the potential sources of microbiological aerosol dispersion; these are the patients with whom we are concerned in the practice of dentistry. The oral cavity, with its rich supply of micro-organisms, is the operating area in which the dentist combines his skill and knowledge in using fine instruments to perform a health service.

Kazantzis (4) reported his observations on bacterial contamination of room air during use of the air-turbine handpiece and conventional handpiece. Air was drawn from the region of the operator's face through a slit sampler containing a rotating blood-agar plate and samples were collected before, during and after each dental procedure employing each handpiece. Although tooth debris was observed on the appropriate segments of the agar plates indicating periods of drilling, Kazantzis observed no significant difference in the amount of deposit between the two techniques.

It was the purpose of our first study (5) to compare aerosols generated by conventional and air-turbine handpieces while cutting tooth structure previously inoculated with a known and easily recognizable bacterium rather than utilizing the patient's flora. Also, because airborne infectious particles may gain access to susceptible tissue by virtue of their aerodynamic dimension (6), we wanted to observe colony concentration and classify the sizes of *Serratia marcescens* containing droplets according to the enumerating and sizing capabilities of the air sampler used.

The aerosols were collected with an Andersen Air Sampler* using nutrient agar plates within the collection chamber of each of the six stages.

Since these first studies were conducted in a large dental clinic, it was necessary to surround the operating area with 6-foot divider screens to minimize the flow of undesirable air currents. A duplicate operating area was established in another section of the clinic so that conventional handpiece preparation could be done in one section and the air-turbine preparation in another. The reason for this location change was to minimize collection of organisms associated with the first sample period.

After administration of the anaesthetic, a rubber dam was placed on the tooth to be treated, the tooth was cleansed with 70 per cent ethyl alcohol, and the operating area dried. The air sampler was loaded with agar-filled Petri plates and placed in position but with the intake orifice covered. The isolated tooth was inoculated with 0·2 ml. of the suspension of *Serratia marcescens,* the air-sampler orifice uncovered, and the motor started, the timer started, and the low-speed handpiece activated. The bur was actually cutting tooth structure during each three-minute sampling period. This procedure was repeated in the other duplicate area but with the air-turbine handpiece and for 15 subsequent paired sequences.

Analysis of air-sampler data revealed that the conventional handpiece, when operated at a speed of 2000 rpm, produced very little air contamination, whereas its air-turbine counterpart, operating at speeds fifty times greater, generated considerable aerosol. Almost 85 per cent of the total mean counts with the air turbine occurred on the last three stages of the air sampler. According to Andersen (7), particles collected on these three stages represent mean diameters of 3·5 μ or less. Riley and O'Grady (6) emphasized that approximately 40 per cent of inhaled particles in the

* Andersen Air Sampler, Model 603, Andersen Samplers and Consulting Service, Provo, Utah.

2–3 μ range will reach the alveoli. Only 47·0 per cent of the total mean counts occurring as a result of the use of the conventional handpiece had a mean diameter of 3·5 μ or less. Thus, almost twice as many droplet nuclei capable of penetrating the alveoli were produced by operating of the air turbine.

Inadequacies of working in a large clinic area with portable screens as the only device to separate the experimental areas made it necessary to establish a dental operatory in which ventilation and air dilution could be controlled to provide more equivalent conditions for assaying the produced aerosols (8).

A room measuring 10 by 12 feet with a 15-foot ceiling was partitioned to provide a operatory of 8 by 10 feet with a 9-foot ceiling. The walls and ceiling were plastered and painted with two coats of a water-resistant paint. After providing a drain, the entire floor was covered with ceramic tile. Aluminium frame windows were caulked and sealed, and the entrance door was fitted with a self-sealing closure. An air conditioner and its ducts, an exhaust fan, and an air compressor were installed in the crawl space above the ceiling. As determined by the use of smoke tubes, the room was under positive internal pressure.

The essential furnishings within the operatory were a dental chair, unit, and a hand sink. The dental chair was positioned so that the headrest bracket was located in the centre of the room. The dental unit was equipped with a standard dental engine* and an air-turbine control unit and handpiece.† A 3- by 4-inch plaster block that could be clamped in position between the headrest and chair back was used as a base on which a freshly extracted third molar could be mounted. The tooth was positioned by pushing the crown into a precut hole in the side of a 4-inch diameter plaster bowl. The bowl was then placed on the block with the opening of the bowl positioned in a manner similar to the open mouth of a patient. The mounted tooth was isolated with a rubber dam. A molar clamp and a Young's rubber dam frame maintained position of the rubber dam. After 0·2 ml. of a suspension of *Serratia marcescens* was placed on the tooth, a Class II cavity was prepared, using a $33\frac{1}{2}$ inverted cone bur in the air turbine handpiece and a 35 inverted cone bur in the conventional handpiece. These bur handpiece combinations were utilized in an attempt to remove approximately the same amount of tooth structure. To determine aerosol decay, three air samplers were placed in a straight line at distances of 10, 20 and 30 inches from the tooth, with the intake orifice of the samplers parallel to the occlusal surface and perpendicular to the buccal surface of the mounted tooth. A Petri dish containing nutrient agar was placed to the side of each air sampler. All air-sampling devices were exposed simultaneously.

The determination of aerosol dispersion was accomplished by locating one air sampler in the same position as the first sampler in the aerosol decay studies. The other two air samplers were located 8 inches to either side of the intake orifice of the first sampler and 10 inches from the tooth. Open petri dishes were placed on either side of and between the air samplers and also immediately in front of the

* Ritter Tri-Dent Model B, Rochester, N.Y. † Borden Airotor, Model B, Ritter Co., Rochester, N.Y.

tooth. Placement of air-sampling equipment with relation to the mounted tooth was constant in all procedures.

Composite data obtained from 17 experiments (Figure 1) demonstrate that the number of droplet nuclei containing *S. marcescens* decreased rapidly as the sampling distance from the mounted tooth increased. The lung penetration ratio (LPR), which is our method of expressing the ratio of the number of bacterial aerosols approximately 5 μ or less in size to the total number of bacterial aerosols collected, increased with the sampling distance.

A somewhat similar situation was observed when the sampling locations were on an arc in front of the mounted tooth (Figure 2). It is interesting to note that the operator influences to a great extent the distribution of aerosols as collected on this arc. Although a left-handed dentist has not been used to demonstrate operator position influence it is rather apparent from these results that such influence most likely does exist. In any event, these results demonstrate that the operator is exposed to large numbers of droplet nuclei, of which at least one-fifth have the capacity of penetrating the alveolar spaces when the air-turbine handpiece is operated without water spray.

Amost identical results were obtained when a group of patients were used as subjects. The patients were evenly divided by sex, varied in age from 14 to 46 and had DMF (Decayed-Missing-Filled) Indexes from a low of 12 up to a high of 25. Thus the experimental model using a mounted tooth and indicator organism was entirely adequate in demonstrating the problem.

All of these studies demonstrated a need for evaluating the various factors of air turbine operation. Questions of considerable importance have been the influence of particle size of water spray, bur shape, construction, material, and size, together with the combined effect of handpiece pressure and water flow rate (9). To investigate these problems a Royco Model 200 Particle Counter* was employed to monitor the aerosol cloud during carefully defined blind replications of tooth-cutting procedures on freshly extracted and mounted human molars, using the Borden air-turbine handpiece, while operating in the dental operatory as previously described.

The essential furnishings of the operatory as shown in Figure 3 were the same as previously described with one exception that the dental chair was replaced by the Royco particle counter. Additional appurtenances in the operatory included a Mayo bracket table, a 50-ml. calibrated glass cylinder, a stop watch, and a freshly extracted lower first molar that was previously weighed and embedded in a plaster block so that only the clinical crown was exposed to the environment.

The Royco particle counter, which has the capacity to count and size particles between 0·3 and 10 μ over 15 selective channels, was used during tooth-cutting procedures to assess the aerosol cloud. Aerosol particle counts of six different size ranges were made by activating the appropriate channels switches.

* Royco Instruments, Inc., Menlo Park, California.

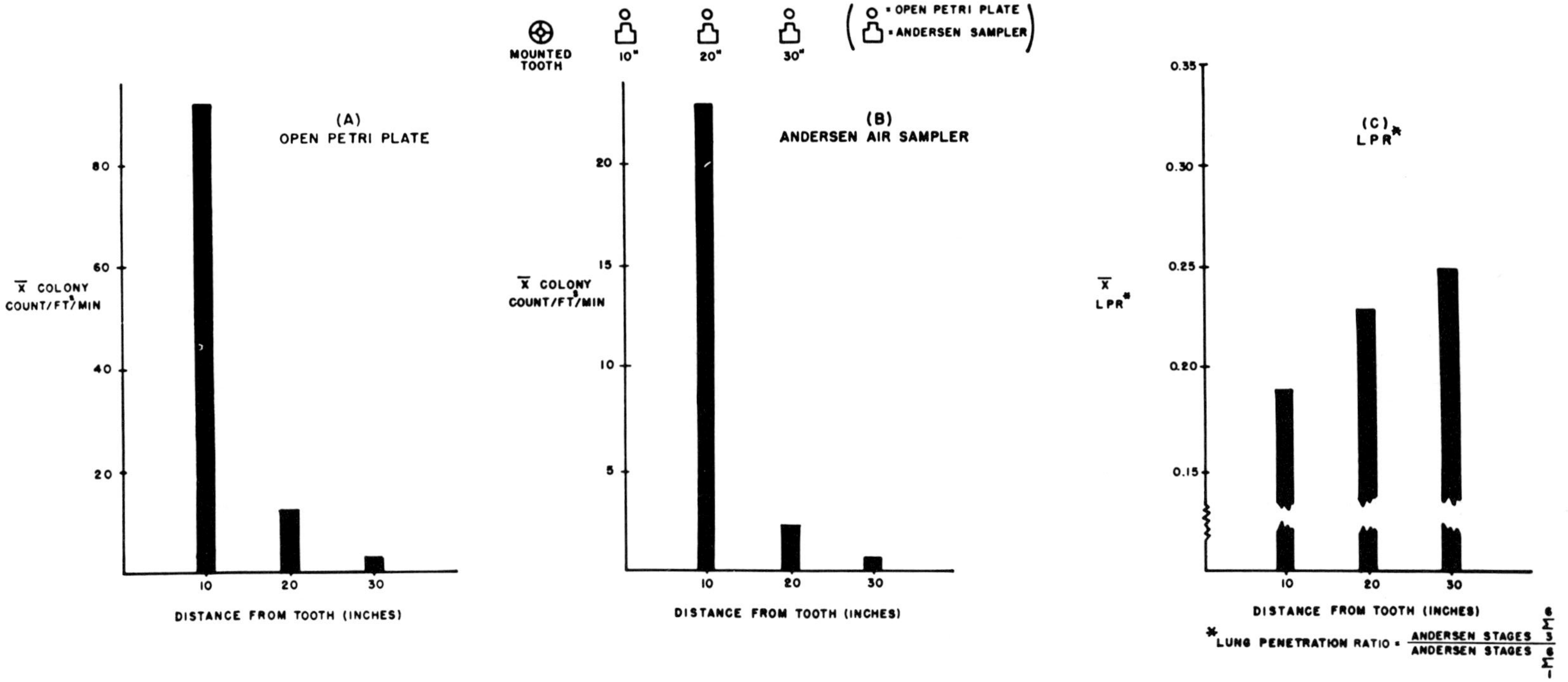

Figure 1. Decay of aerosol containing *Serratia marcescens* generated by air-turbine handpiece.

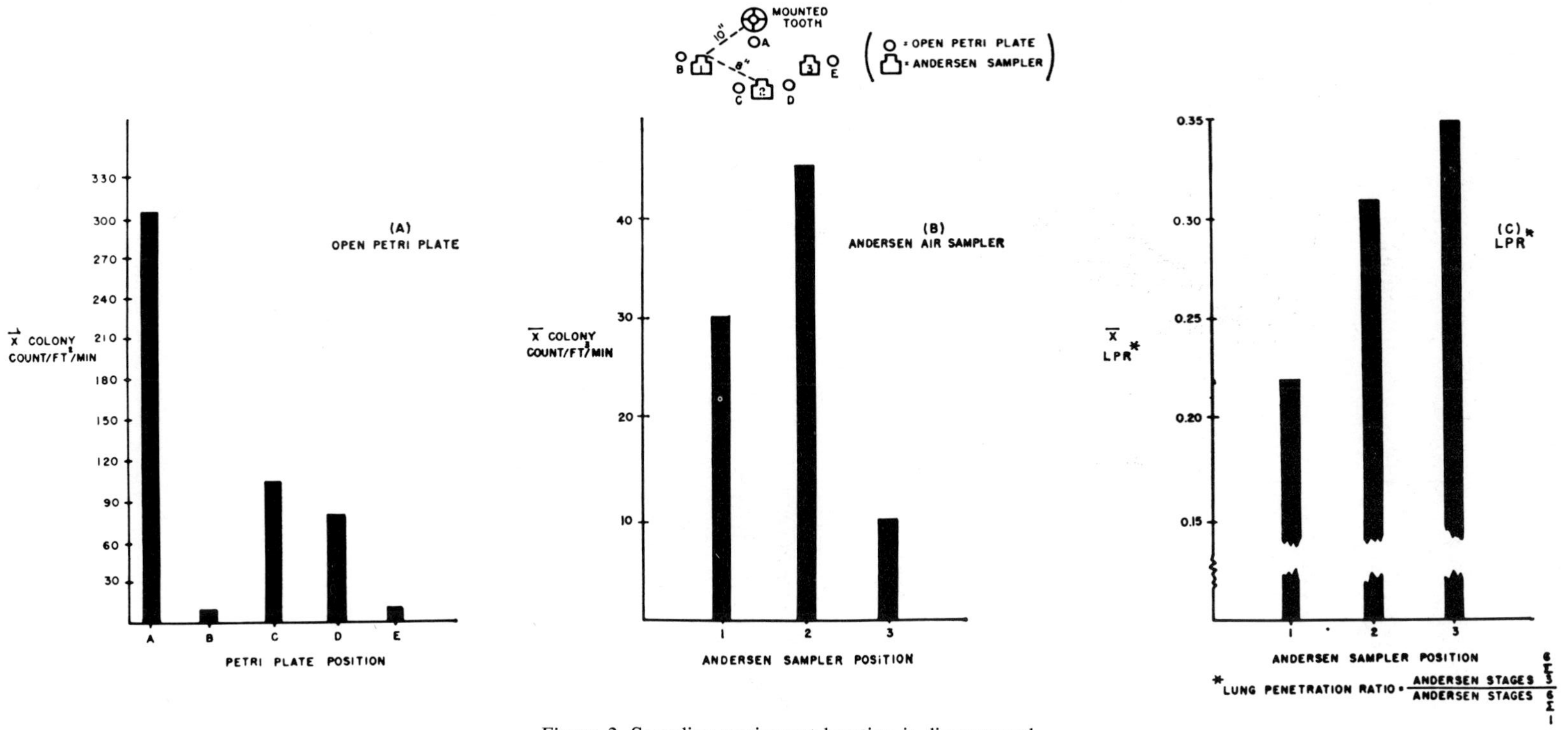

Figure 2. Sampling equipment location is diagrammed.

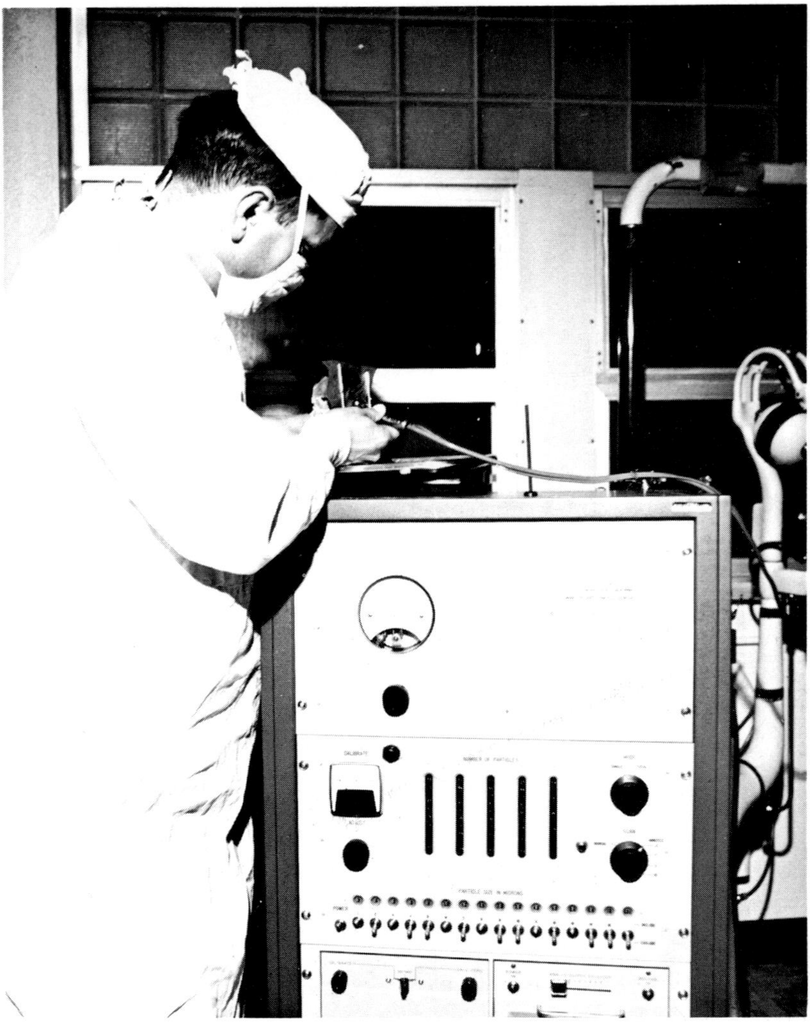

Figure 3. Using Royco Particle Counter while cutting tooth structure.

Preliminary experiments aided in establishing the optimum sampling period. It was observed that six distinct buccal-lingual cuts averaging 5 mm. in depth could be made in the occlusal surface of the mounted tooth. Each cut could be accomplished using light pressure and maximum turbine speed during a two-minute test period. The particle counter was programmed to sample each of the six selected channels for 0·3 minute per channel and the device required 0·2 minute to recycle from the last to the first channel.

Three bur shapes selected for the experiments included cylinder, inverted cone and round, all in comparable size and of diamond and carbide construction. The air-turbine control-box water-flow needle valve was calibrated to permit the operator to accurately set the rate at no flow, 27 ml. per minute and 52 ml. per minute. Thus there was a total of 18 different combinations.

A previously weighed and mounted molar was placed on a Mayo stand and brought into proper relationship with the intake orifice of the particle counter. The Mayo stand was adjusted to permit the occlusal surface of the mounted tooth to be parallel to the intake of the particle counter and 10 inches from the buccal surface. The statistically selected experimental combinations for the six cutting procedures on the mounted tooth were noted and the proper bur and waterflow rate for the first of six cuts was prepared.

The particle counter was allowed to cycle through the six selected channels five times, and on the start of the sixth cycle the handpiece was activated. At the start of the seventh cycle the handpiece was deactivated and the cutting combination for the second cut was prepared. The particle counter was allowed to cycle through the selected channels for 10 minutes (five cycles), and on the start of the 11th minute (sixth cycle) the handpiece was activated and the second two-minute cut made on the mounted tooth and then a final five-cycle background count was recorded.

Each experiment was carried out in accordance with a replicate plot design. Any error in the procedure, sequence, or discrepancy noted on final calibration check of the Royco particle counter disqualified the entire block of experiments.

The experiments were designed to compare the number of airborne particles observed under 18 different experimental conditions (10). These conditions were formed by considering all combinations of the following: two bur materials; three water-flow rates; and three bur shapes.

The analyses of variance revealed the following:

1. Flow-rate differences were statistically significant at $P < 0{\cdot}01$ for all channels. It appears that the low water-flow rate (no water) generally leads to lower particle counts while there is little difference between the particle counts when flow rates of 52 ml. or 27 ml. per minute were employed. Miller *et al.* (11) has also reported this observation.
2. No statistically significant difference at the 0·05 level in the particle sizes of the aerosol cloud generated during the cutting procedure could be found as a result of using cutting burs of different shapes, i.e. round, inverted cone or cylinder.
3. There was no statistically significant overall difference between the carbide and diamond bur material except for the $0{\cdot}3$–$0{\cdot}4\ \mu$ and the $0{\cdot}5$–$0{\cdot}6\mu$ particle level when $P = < 0{\cdot}05$.
4. None of the interactions were statistically significant at the 0·05 level; therefore, no dependency.

It was further observed that with all cutting combinations 99 per cent of the aerosol cloud consistently contained particles 5 μ or less in size. This result confirms earlier observations obtained by counting and sizing particles which contained bacteria.

Micik *et al.* (12) conducted similar studies in a test chamber designed and constructed to provide a close system with minimal background air contamination. Filtered air flow was adjusted by means of flowmeters and the air samplers were connected to the system by means of a manifold. They reported that a reduction of bacterial aerosols in excess of 99 per cent could be accomplished by using a high-velocity suction device either held by an attendant adjacent to the aerosol generating instrument or by placing it in the patient's mouth.

Another dental treatment device which, during normal operation, produces considerable aerosol is the cavitron. This is an ultrasonic unit that incorporates a water coolant spray. In studies conducted by Mazzarella and Hancock (13) aerosols were found to contain greater than 5000 bacteria per cubic foot of air and 60 per cent of the particles were in the size range of 1 to 5 microns.

All of these studies demonstrate that there is a problem when using super-speed dental handpieces and the ultrasonic oral prophylaxis instrument. Some reduction in the quantity of aerosols expelled into the operatory environment can be effected by operating without a water flow and at the same time using a high-velocity suction device. On the other hand, one could use an extremely high water flow to flood the field which then aggregates the small particles, but then visibility is reduced and excessive flooding of the operating field occurs. Both of these situations if practiced will add some operational difficulties and will thereby reduce their possibility of use by the practitioner. Many dentists refuse to treat patients using air-turbine handpieces without a water coolant spray. They maintain that the underlying damage to dental tissue due to the heat generated is sufficient cause to expose themselves to whatever infectious aerosols are produced by their procedures. Studies are needed to determine the optimum design of a dental operatory and appropriate protective devices for the dentist.

Dental operatories are normally simple enclosures consisting of a chair, instrument cabinet, lavatory, X-ray unit and an overhead light. A large space is not required and most operatories are an approximate 8-foot cube. The design of air-flow patterns that would allow for a downwash and exhaust through the peripheral baseboard might solve many of the problems. Another area that should at least receive passing consideration would be the design of a protective face mask or helmet for the dentist. However, it should be noted that most dentists feel that such a device may prove too clumsy and alarming to the patient even though its use has been adequately explained.

Comments have been made in the past that all air-turbine aerosols studies have been based on bacterial recoveries and that they should be extended to include viruses. This seems to us to be a waste of investigator time and talent, and results would be only academic. Aerosols have been produced and characterized; therefore, it would seem that future efforts should be turned to design of operatories and personnel protection. In the design of operatories the main difficulty will be to construct an efficient air-handling system and will not appreciably increase the cost of construction. In other words, risks of disease that so far are mostly circumstantial are more readily accepted than economic risks.

Thus it would seem to us that future investigations of dental aerosols should be concerned with demonstrating the risk of infection by epidemiologic models and designing an operatory based on the most current methods of air handling and distribution.

SUMMARY

Studies of the authors and a number of other investigators have demonstrated that through normal operating procedures the air-turbine handpiece produces a considerable aerosol which contaminates the dental operatory with micro-organisms from the patient's oral cavity. It would now seem more appropriate to turn attention to methods of reducing the probability of exposure to infectious agents by designing model air-handling dental operatory systems. It is believed that appropriate room design can be recommended based on the results of these dental aerosol investigations and providing that the cost is not excessive.

REFERENCES

1. Bernier, J. L. and Knapp, M. J. (1959). Methods used in evaluation of high speed instruments and some results, *Oral Surg., Oral Med., Oral Path.*, **12,** 234–252.
2. Bernier, J. L., Knapp, M. J. and Boyers, R. C. (1960). Pros and cons on high speed rotary instruments, *Dent. Prog.*, **1,** 51–55.
3. Flick, L. F. (1910), Tuberculosis preventive measures for the dentist and his clientele, *Dent. Cosmos.*, **52,** 1332–1335.
4. Kazantzis, M. G. (1961), Air contamination from high speed dental drills, *Proc. roy. Soc. Med.*, **54,** 242–244.
5. Madden, R. M. and Hausler, W. J., Jr. (1963), Microbiological comparison of dental handpieces. 1. Preliminary report, *J. dent. Res.*, **42,** 1146–1151.
6. Riley, R. L. and O'Grady, F. (1961). *Air-borne Infections: Transmission and Control*, Macmillan Co., New York.
7. Andersen, A. A. (1958). A new sampler for the collection, sizing and enumeration of viable air-borne bacteria, *J. Bacteriol.*, **76,** 471–484.
8. Hausler, W. J., Jr. and Madden, R. M. (1966). Microbiological comparison of dental handpieces. 2. Aerosol decay and dispersion, *J. dent. Res.*, **45,** 52–58.

9. Madden, R. M., Hausler, W. J., Jr. and Leaverton, P. E. (1969). The air turbine handpiece: and analysis of clinical variable, *J. dent. Res.*, **48,** 341–345.
10. Li, C. C. (1964). *Introduction to Experimental Statistics*, McGraw-Hill Co., New York.
11. Miller, R. L., Burton, W. E. and Spore, R. W. (1963). Aerosols produced by dental instrumentation, *Proc. First Intern. Symp. Aerobiol.*, Berkeley, California.
12. Micik, R. E., Miller, R. L., Mazzarella, M. A. and Ryge, Gunnar (1969). Studies on dental aerobiology: 1. Bacterial aerosols generated during dental procedures. *J. dent. Res.,* **48,** 49–56.
13. Mazzarella, M. A. and Hancock, J. G. (1968). Dissemination of bacterial aerosols by an ultrasonic dental prophylaxis unit. Abstract in *Bacteriological Proceedings*, M31.

Introduction to intramural spread of infection

Philip S. Brachman

National Communicable Disease Centre, Atlanta, Georgia

There are two main routes by which micro-organisms are spread within hospitals: the contact and the airborne routes. The contact route involves direct contact, i.e. from person to person; indirect contact, i.e. from the inanimate environment to the person; and by means of droplets, such as contaminated aerosols from humidifiers and nebulizers. The contact route is the most common means by which infection is spread within hospitals.

The relative importance of the airborne route in the transmission of nosocomial infection has long been debated. Assessments have been based on experimental studies, as well as on data from studies of the natural occurrence of disease. Respiratory diseases such as influenza and pulmonary tuberculosis are readily accepted as examples of infections spread among hospitalized patients by the airborne route. However, in considering other diseases, such as staphylococcal wound infections, the route of transmission in a particular case is not as discernible because of the multiplicity of possible routes of infection and the ubiquity of the organism.

Similarly, the route by which a person becomes colonized with virulent organisms is not always clearly defined, but the airborne route is frequently involved, as in nasal colonization with staphylococci. The carrier may subsequently be the source of infection for himself or for other susceptible individuals. Why some carriers disseminate organisms and others do not is not totally clear. It has been shown that the frequency with which disease follows airborne dissemination is generally low.

In addition to the human reservoir, there are inanimate sources of airborne organisms. For example, outbreaks of disease have followed the resuspension of contaminated dust (salmonellosis, tetanus).

The presence of micro-organisms in the air is only one aspect related to the subsequent occurrence of disease; virulence and quantity or organisms, size of airborne particles, host-susceptibility, and site of implantation are all significant determinants of disease.

Transmission of staphylococci in newborn nurseries

Edward A. Mortimer, Jr.

Department of Pediatrics, University of New Mexico, Albuquerque

These studies were devised in an effort to clarify routes of transmission of staphylococci among newborn infants in nurseries. Early in planning it became evident that multiple sources and possible routes precluded the derivation of definitive epidemiologic data from simple culture surveys, even with phage typing. Accordingly in these studies nursery practices were modified to limit exposure of newborn infants to only one or two possible sources or routes of transmission, or to 'block' a single route. Employing such a study design it was anticipated that epidemiologic data susceptible to simple analysis might be obtained.

Multiple studies were performed in which various possible modes of spread were examined. Staphylococci were identified by standard bacteriophage typing techniques.

The results indicated that transmission may occur in several ways at different rates. The airborne route appeared to be of minimum importance, as might be suspected from the fact that the umbilical stump is the usual site of initial colonization. Fomites were shown to be capable of transmission, but the degree of contamination required far exceeded that which would exist following the most elementary laundry techniques. Instead, these studies indicated that colonization of newborn infants was usually a consequence of the transfer of organisms to infants by the hands of personnel. This appeared to occur in two ways: transfer of a personnel-carrier's organisms to an infant by the carrier's own hands ('primary physical transfer'), or transfer of bacteria from a colonized infant to another infant by the hands of a nursery attendant who was not a carrier in the usual sense ('secondary physical transfer'). Studies of interruption of transmission indicated that simple handwashing was usually sufficient to interrupt this mode of spread.

Studies on the epidemiology of postoperative staphylococcal infections in a thoracic surgical unit

G. Laurell and **G. Lindbom**

Department of Clinical Bacteriology, University of Uppsala, Sweden

---

The aim of the study was to analyze *when or where* did the patients acquire disease-producing staphylococci, *which* were the sources, and *how* were the bacteria disseminated with special reference to the importance of air.

Of the 375 patients studied 72 (19·2 per cent) developed infections after intrathoracic surgery. *Staph. aureus* was isolated from 37 (9·9 per cent).

In the operating theatre, a total of eight infections (22 per cent) were classified as exogenous and theatre-acquired. The theatre was reasonably well ventilated and discipline was good. The median staphylococcal fallout during operation was approximately 0·02 colony per dm.^{2}/5 hrs. The operating team of 12 regular members included seven persistent nasopharyngeal carriers. One surgeon with psoriasis was a heavy disperser and the ventilation was sometimes tried too severely, the fallout occasionally rising to 5 colonies per dm.^{2}/5 hrs. Possibly due to low virulence, staphylococci dispersed by the surgeon caused only two infections, most likely by direct fallout from the air or by a process involving both airborne dissemination and contact-transfer. A more dangerous source was a theatre sister who developed a paronychia. Though never a heavy disseminator, she caused five serious infections, most probably by contact-transfer.

The endogenous route probably accounted for infections in 12 patients (32 per cent). All were nasopharyngeal carriers of the same strain at operation that later caused the postoperative infection.

Finally the importance of the intensive care ward, where 13 patients (35 per cent) acquired infections, was also studied. The discipline there was also good, but control measures were hampered by lack of mechanical ventilation and insufficient isolation facilities. Therefore the staphylococcal fall-out fluctuated widely, depending upon whether dispersers, usually tracheotomized patients, were present. When that was the case, the fall-out ranged from 6 to 19 colonies per dm.^{2}/5 hrs. The result was massive environmental contamination in the *whole* unit, and frequent colonization of exposed patients, particularly if tracheotomized. There was a good correlation between weekly counts of staphylococci and exogenous, ward-acquired infections. Thus, in the presence of a disperser, staphylococci were spread by air, causing infections either by direct fall-out into exposed tissues, or by colonizing patients in the nasopharynx and secondarily by contaminating their wounds, or by a multistage process involving heavy contamination of objects from which they reached susceptible sites by contact-transfer.

The spread of Pseudomonas infection

G. A. J. Ayliffe

Summerfield Hospital, Birmingham

Sources of *Pseudomonas aeruginosa* in hospital may be environmental or human. Environmental sources are of special importance with this organism, since it survives well in moist conditions and will multiply in the presence of a minimal amount of nutrient at room temperature. *Ps. aeruginosa* is commonly isolated from moist areas or equipment in the hospital environment, but infection from many of the possible sources (e.g. mops, sinks and drains) appears to be infrequent. Epidemics of environmental origin are usually dependent on close contact of the source (e.g. contaminated catheters, eyedrops, shaving brushes, respiratory apparatus) with a susceptible site on the patient.

Self-infection may occur from the faeces or upper respiratory tract, but cross-infection from infected lesions or colonized sites is of greater relevance in burns, intensive care, urological and premature baby units. Transfer of organisms on the hands of the staff is probably the most important mode of transmission in these situations.

Ps. aeruginosa dies rapidly on drying, although small numbers of survivors may remain alive in a dry state for long periods; larger numbers may survive if protected by organic matter. The organism can rarely be isolated from the air, and there is little evidence to suggest that airborne transfer is a major route of spread of infection.

Spread of *Pseudomonas aeruginosa* (*Pyocyanea*) and *Escherichia coli*

R. A. Shooter

Department of Bacteriology, St Bartholomew's Hospital, London

Some infections with *Ps. aeruginosa* and *Esch. coli* are caused by organisms carried by the patient. Current work is designed to trace the routes by which these two organisms spread in hospital.

Extensive surveys have shown that during their stay in hospital, surgical and medical patients acquire fresh types of *Ps. aeruginosa* and *Esch. coli* in their intestines. This was at first thought to be the result of patient to patient transfer. As yet we have failed to detect evidence of airborne spread and we have little to suggest that these bacteria are transferred by articles in common use.

There are, however, reasons for thinking that, at any rate for the patients we have observed, a proportion of the freshly acquired organisms come from food.

The evidence for airborne infection

W. C. Noble

Department of Bacteriology, St. John's Hospital for Diseases of the Skin, London

Almost every pathogenic bacterium and fungus has been recorded as present in the air of a hospital or institution on some occasion. Staphylococci, streptococci, clostridia and candida have been isolated from the air of hospital wards and yeasts and dermatophyte fungi from the air of barracks. Not surprisingly, therefore, a number of reports can be found of infection due to direct airborne transfer. Respiratory pathogens such as *Mycobacterium tuberculosis* and *Coccidioides immitis* are well documented, but there are also reports of infection following surgery where the organisms were carried and dispersed by a person in the operating theatre who had no direct contact with the patient.

However, most published reports of cross-infection do not indicate this direct route of spread as the most important. Patients in hospital wards frequently become nasal carriers of strains of staphylococci with which they subsequently develop surgical wound sepsis. In these patients the route of spread is probably 'airborne' from source to patient's nose and 'direct' or 'contact' from the nose to the lesion. In patients with skin diseases it may merely be necessary to contaminate the environment of the patient rather than effect colonization before skin 'sepsis' appears. In all these infections some form of potentiator most likely exists which enables sepsis to become manifest. Such a potentiator may be necrotic tissue or debris in a surgical lesion, or pre-existing eczema or chafing in skin disease.

It is difficult or impossible to disentangle the various factors for adults, but in newborn babies only about 10 per cent of the acquisition of organisms appears to occur by the airborne route. High air counts of a particular pathogen in the air of a ward may act as a marker of dissemination and thus potential cross-infection, but it must not be assumed that low air counts indicate absence of cross-infection.

Intramural spread of infection, airborne or not?

Résumé: presented by **K. C. Winkler**

After a general introduction on possible routes and modes of intramural spread and the importance of surveillance it was argued that (1) every organism has its own sources and routes, (2) in every surrounding, even for a specific organism, the relative importance of each route may be very different, (3) every epidemiological situation is unique in space and time, and that, consequently, generalization is dangerous. These principles were well illustrated by the following instances.

(*a*) *Spread of staphylococci in a nursery*. It was shown that a nurse spending her morning by reading aloud to the babies did *not* transmit the staphylococci carried in her nose. However, when the baby was protected in an isolator against other infections and handled with bare hands by the nurse, staphylococci were generously transmitted. The main routes seemed to be by primary physical transfer (nurse's nose—skin—baby) and secondary physical transfer (baby A—hands of nurse—baby B). A good handwashing routine was effective but cost valuable time. The airborne route seemed to be of minor importance and only responsible for 10 per cent of infections. This seemed to tally with the dose that the baby would inhale taking Shinfield's data about infective doses as a guide. It was again stressed that the uncolonized skin of the newborn is very sensitive and easily provides board and lodging for travelling staphylococci.

(*b*) *Staphylococcal infection in thoracic surgery*. It is well known that the prolonged thoracic operations often on patients with low resistance are liable to give rise to infections. In the specific instance discussed 19 per cent of about 400 patients became infected; 9 per cent of these were due to *Staph. aureus*. About 30 of the staphylococcal infections were endogenous. A few could be traced to a surgeon with psoriasis who was a heavy disperser; some other infections were due to a nurse with a staphylococcal lesion. About 35 per cent were obtained in the recovery room. The importance of a tracheostomy as an easily infected target, as well as an effective direct spray, was stressed.

In connection with staphylococcal infection it was shown that staphylococci which first enter the nose generally appear later on the skin, whereas staphylococci acquired on the skin do not necessarily transfer to the nose. The number of scales shed from the skin can be counted and varies widely. From the axilla, larger and thinner scales are shed. Whether, by floating over long distances, these scales are a method of large-scale dissemination is unknown. The increase of hospital infection due to gram negatives was touched on only briefly, mentioning the increase of susceptible patients as one possible cause for the more serious cases. The closer nursing attention needed for such infections was mentioned as a second factor.

The specific case of pseudomonas infections is complicated by the fact that this organism can multiply in the environment. Though it dies rapidly on drying, part of the population can survive especially if protected by organic matter.

Many infections seem to be environmental when respiratory apparatus, catheters or eyedrops loaded with a sufficient dose are carried to a susceptible site on the patient. When strains from sinks, shaving brushes, etc., are typed serologically with pyocines or with phages, the environmental strains are often different from the epidemic strain. It was stressed that the presence of an organism does not mean that it is a source. When a lesion exists in a ward pseudomonads are also found in the hands of staff, which could lead to further spread.

The contribution of the airborne route seems small except in the cases of tracheostomy and, of course, burns.

Self-infection with pseudomonads from the gut seems possible. It was shown, however, that after growth the organisms are excreted for a short time only and seem not to multiply extensively. By careful analysis and typing the same seemed true for patients who picked up strains in hospitals, often with food as the source.

Serological typing of *E. coli* strains in faeces point in the same direction. Strains can be picked up but often seem to be excreted for one or two weeks only. People carrying up to five serotypes are not exceptional.

From these instances it was argued that airborne transmission is not a major route of intramural spread of bacteria.

To prevent cross-infection in hospitals much can be done by simple means and by strict routine. It was realized that the law of diminishing returns is affecting the elimination of the remaining infections. The general message seemed to be that hospital epidemiologists and aerobiologists should join hands in providing evidence and stating clearly what is necessary or desirable in hospital design and facilities.

Dry-heat sterilization under high vacuum

William S. Miller

Dickinson and Co., Raleigh, North Carolina

A unique cabinet sterilization program is being carried out in the Lunar Receiving Laboratory at the Manned Spacecraft Center, Houston, Texas. The cabinets are being used to provide high vacuum conditions during initial handling and examination of returned lunar samples. The vacuum cabinet complex must be sterile to prevent biological contamination of lunar samples by terrestrial contaminants. To achieve this condition, a bakeout process was developed which employed dry heat at a vacuum of about 1×10^{-6} torr. The method of sterilization was developed from initial observations on the bakeout process under high vacuum which indicated, in a 44 hour cycle, complete sterilization of 1×10^{6} *B. subtilis* spores at an average temperature of 85°C and a maximum temperature of 89°C. Subsequent trials with a temperature exposure time of 24 hours indicated complete kill of 1×10^{6} spores at an average temperature of 108°C with a maximum of 115°C. Additional investigations of the apparent enhancement of sterilization by high vacuum indicated that *B. subtilis* showed no significant reduction in count after 20 hours' exposure to 90°C at atmospheric but a two-log reduction in count after 20 hours' at 90°C under vacuum. At 110°C there was a two-log reduction at atmospheric pressure and sterilization 1×10^{6} spores under vacuum. With the anaerobe *Clostridium sporogenes*, it was found that there was a greater sensitivity to heat with a two-log reduction at 90°C for two hours at atmospheric pressure. With the application of vacuum, there was an additional reduction in count at two hours of two logarithms. Possible explanation of these effects is given by published work which suggests that lower D values are obtained with decreasing spore water activity values. It might be noted that a water activity of 0·9 corresponds to a moisture content in spores of about 50 per cent while a water activity of 0·1 indicates a moisture content in spores of about 6 per cent. It was observed by Bocchino that when air was passed over *B. subtilis* at 110°C there was a marked effect of air moisture content on D values. A D value 37 minutes was obtained when dry air was passed over the spores. On the other hand, a D value of 660 minutes was obtained when the air contained 0·16 lb of water per lb of gas. These results then strongly suggest that enhanced sensitivity of spores to dry heat in these studies is the result of dehydration resulting from the vacuum condition.

Tests for the germicidal efficiency of ultraviolet installations in the laboratory

W. K. Thomson, Claudette Malysheff and **D. C. O'Connell**

Defence Research Establishment Ottawa

Many laboratories, pharmaceutical firms, meat storage plants, hospitals, school rooms and even some home air conditioners and clothes driers are now equipped with ultraviolet lamps. Their value for destroying airborne bacteria and viruses has been well demonstrated, if the germicidal lamps are operating efficiently. Since the latter is by no means a certainty in all cases, the lamps should be examined at regular intervals.

Visual examination is not dependable, since lamps which have been in use longer than their rated life may *appear* to be germicidal because they are illuminated. Nor is strict adherence to a policy of replacing lamps after 7500 hours' rated life a good practice, since lamps used intermittently for short starts and lamps used in cold rooms have a reduced life.

A survey of lamps in our laboratories demonstrated that lamps should be examined routinely and that germicidal installations should be standardized to assure that they inactivate bacteria and viruses. The survey included ceiling lamps, lamps in air ducts and air locks, all of which share in the control of pathogens which might become airborne and in the reduction in the number of air contaminants that interfere with laboratory procedures. Since the choice of method was found to be influenced by the number of lamps and the type of installation, several methods of testing lamps were investigated and compared.

Safety cabinets

G. Thomas

Microbiological Research Establishment, Porton

Infection of laboratory personnel occurs most commonly by the airborne route. Totally safe handling of pathogenic organisms which cause disease in this way is only possible by conducting virtually all manipulations in specially designed handling cabinets. Long experience in this field at the Microbiological Research Establishment has shown that these cabinets must embody certain essential design features in order that they may be safe under all practical conditions of use.

Many commercial models available have been examined and found to be lacking in one or more essential features, leading to possible hazardous situations arising when pathogens are handled in them.

The main requirement is that any aerosol of infective organisms which arises during manipulations must be contained within the cabinet. To prevent escape, cabinets are usually ventilated to ensure that air constantly passes into the cabinet via the glove ports, loading doors, etc., when in use. Air flow rates are critical and must be sufficiently high (100–200 linear feet/minute) to prevent reflux into the laboratory when hands are withdrawn, etc. Exhaust air from the cabinet is most effectively dealt with by filtration, using filters which have been thoroughly tested in situ. Under these conditions air may be returned directly to the laboratory. Decontamination of the interior surfaces of the cabinet and of the apparatus it contains is best achieved by fumigation with formalin. Consequently the cabinet should be capable of being totally closed and should have built in the necessary equipment for dispersal of formaldehyde vapour.

Laboratory safety

Résumé: presented by **H. M. Darlow**

Dr. W. S. Miller opened the seminar with a paper on heat sterilization under high vacuum. He described experiments in which heavy inocula of *B. subtilis* spores were killed at temperatures ranging from 85 to 115°C at a vacuum of about 1×10^{-5} torr. The time to achieve 100 per cent kill ranged from 24 to 44 hours. Further experiments with *C. sporogenes* indicated that this organism may be considerably more sensitive. He produced evidence to suggest that sensitivity to dry heat at modest temperatures was a function of the diminished water content of the spores. In the subsequent discussion, however, it was suggested that the kill might not occur until the inocula were removed from high vacuum, and that sudden rehydration or exposure to oxygen might be the lethal factors. Carrying out sterility tests in a nitrogen atmosphere would help to elucidate this aspect. The possibility of super-dormancy was also raised.

Whilst these findings were considered remarkable and worthy of further investigation, the prolonged killing time at present limits their practical application, though it is to be hoped that the principle can be used for the sterilization of heat-labile materials.

Dr. D. C. O'Connel then gave a paper on practical experiences of U.V.R. as a laboratory safety measure. He described the standard types of system and their maintenance, paying particular attention to the great importance of frequent output monitoring, a procedure often ignored. He gave details of two types of radiation meter and of photographic and bioassay techniques, and compared their respective values in different circumstances. A general discussion of personal experiences, hazards, uses and abuses ensued and the relative merits of various types of U.V. lamp were also mooted.

Dr. G. Thomas then presented a review of published work and personal experiences of a wide range of safety cabinets, stressing the vital importance of maintaining an adequate air through-put both to prevent escape of organisms and to dilute any aerosols present within the cabinet. He gave valuable details of sterilization and testing techniques, particularly in respect of exhaust air filters.

This automatically led to a lively discussion of all aspects of laboratory safety, including disinfectants, fumigation, filter testing, disposal of contaminated material and liquid effluents, immunization, general medical supervision, and staff training.

The range of subjects and the spirit in which the seminar developed suggested that the participants derived considerable benefit, and that similar sessions should be included in future symposia.

Immunological aspects of pathogenesis

S. Gard

Department of Virus Research, Karolinska Institute, Stockholm

---

Coombs and Smith have classified immune reactions as either beneficial or harmful and distinguish four main types of each category. The division into classes on the basis of pathogenic effects is hardly rational, however. The same reaction may be beneficial or harmful depending on the circumstances and may in fact simultaneously act in both directions. For instance, circulating antibodies may seal off a focus of infection and prevent virus dissemination. Yet, if infection persists and virus or viral antigen is continuously being released, antigen-antibody reactions may cause local Arthus' phenomena or nephropathy after deposition in the kidneys of antigen-antibody-complement complexes.

Most studies have been concerned with the state of immunity and with reactions occurring upon re-exposure to an antigen. The above-mentioned four main types of immune reactions appear relatively late in the course of a primary infection and their importance as controlling factors in the acute stage is debatable. Apparently other defence mechanisms are in operation before the appearance of antibodies and immunocytes. Two such mechanisms, both non-specific, are well known: inflammation and interferon production.

Early, specific reactions may, however, also be induced. Studying measles immunity, Gard, Norrby and Lagercrantz detected a possible qualitative difference between the effects of inactivated vaccines and of actual infection. Post-infection serum seems to contain a protective factor not produced after administration of inactivated vaccine and not demonstrable in conventional neutralization tests.

Similar observations were reported by Thind and Price working with arboviruses. They described a 'serum protective factor' (SPF), which appears early in infection, is a thermolabile, specific 7S globulin, and protects cells against infection *in vivo* and *in vitro*. SPF appeared only after actual infection; it was formed even if production of neutralizing antibody was completely suppressed by cytostatic drugs. The phenomena of immunological memory and 'immunological sin' were demonstrable.

According to Huldt, infection with avirulent toxaplasma induces resistance to virulent strains, clearly demonstrable within 12 hours after infection. A thermolabile factor, protecting cells *in vitro* was found in serum long before antibodies appeared. Immunization with killed toxoplasma does not confer protection.

Since measles in children with thymus aplasia usually is lethal, it has been suggested (Burnet) that cell-associated immunity might play a decisive role in the defence against the infection. The remarkable protection afforded by gammaglobulin, indicative of humoral rather than cellular mechanisms, is not however really compatible with this assumption. It may also be pointed out that the histopathology of the rash is that typical of an Arthus' reaction, not that of delayed hypersensitivity.

It appears more attractive tentatively to consider the above-mentioned phenomena as manifestations of a common or similar immunological mechanism of a heretofore unrecognized type. Measles and arbovirus infections as well as toxoplasma, all are intracellular parasites. If SPF is induced only by actual infection, this might indicate that the specific receptors which recognize the antigens and trigger the synthesis of SPF are to be found only in the interior of the competent cells. Obviously these questions merit further studies.

In vivo measure of suppression of cellular immune and humoral antibody response by antilymphocyte globulin

E. Stanbridge, F. T. Perkins, F. Sheffield and **K. Cozens**

National Institute for Medical Research (Hampstead Laboratories), London

---

A most important function of the M.R.C. Division of Immunological Products Control is to ensure the safety of vaccines to the best of our ability, making use of all modern techniques available to us. The development of antilymphocyte globulin (ALG) gave us the opportunity to suppress the immune response of animals so that the oncogenic properties of viruses and tumourogenic properties of new cell substrates may be measured in the absence of the defence mechanisms of the host. It was important, however, to ensure that both the primary and secondary cellular and humoral responses were completely suppressed, and these experiments were designed to illustrate this.

The antilymphocyte globulins were prepared in rabbits by inoculations of mouse thymocytes and the globulins of the rabbit sera were obtained by precipitation with saturated ammonium sulphate. All experiments were done in mice using inoculations of HeLa cells to measure suppression of cellular immunity and survival of mice to tetanus toxin as an index of suppression of humoral response.

The results show quite clearly that both the primary and secondary cellular responses were suppressed, since only those mice receiving ALG had tumours 10 to 14 days after inoculation of the HeLa cells. Similarly the suppression of the primary humoral response was clear cut, since all mice given ALG with a primary stimulus of tetanus toxoid became paralysed when challenged with tetanus toxin. The suppression of secondary humoral response, however, was more difficult to demonstrate. In this case the primary response had to be chosen so that it was sufficient to sensitize but not protect the mice against active challenge. Similarly the secondary dose of toxoid had to be such that it was sufficient to induce a secondary response in sensitized mice but not sufficient to give a primary protective stimulus to unsensitized mice. The results gave an indication that the secondary response was also suppressed, but more work is necessary to obtain unequivocal results. Nevertheless a powerful and useful method of measuring pathogenicity of viruses and cells in laboratory animals has been demonstrated.

Influence of immunologic factors in RS virus disease of the lower respiratory tract

Robert M. Chanock

National Institute of Health, Bethesda, Maryland

Respiratory syncytial (RS) virus is the most important respiratory tract pathogen of early life. In large urban population centres it causes an extensive epidemic every year, easily recognized by a dramatic increase in hospitalization of young infants for serious lower respiratory tract disease. The disease occurs most often in the first few months of life and its frequency then decreases with age, a pattern not seen with other respiratory tract pathogens.

This age–illness relationship resembles the age distribution of passively acquired (maternally transmitted) serum neutralizing antibody (SNA) which infants possess until the sixth/seventh month of life. Infants who develop RS virus disease do not differ in antibody from others. The observation that serious RS disease occurs most often in infants with the highest levels of passively transferred antibody clearly indicates that SNA (presumably IgG) *per se* does not provide effective protection.

Less serious illness follows RS virus reinfection in older children and in adults than primary infection in young infants. Since SNA appears not to be protective we investigated the role of local respiratory tract secretory, 11S IgA, in adult volunteers and found that this type of antibody was the main determinant of resistance.

The most recent experience on the potential role of immunologic factors in RS virus disease was acquired during a series of controlled field evaluations of an inactivated RS vaccine prepared in vervet monkey kidney tissue. The vaccine appeared to be antigenic in infants and young children without detectable SNA. Vaccine-induced serum antibody did not protect against infection; more striking was the unexpected response to infection exhibited by the vaccinees. Nine of 15 (60 per cent) vaccinees 6–23 months of age developed pneumonia and/or bronchiolitis, 5 of them requiring hospitalization, whereas only 4 of 47 (8 per cent) unvaccinated controls developed pneumonia and none required hospitalization. This difference indicated that the vaccine had induced an altered, exaggerated reactivity to infection. Definite evidence of such an altered response was not seen in the older vaccinees, 24–65 months of age. Another study of infants gave similar unmistakable evidence of vaccine-induced altered reactivity to RS virus infection.

The observations which indicated that SNA *per se* did not effectively protect against RS virus disease also suggest that serum antibody may contribute to the development of serious lung damage during infection in infancy. Thus, RS virus bronchiolitis and pneumonia occur most commonly during the first few months of life, when maternally transmitted antibody is at its highest level.

Any hypothesis about the pathogenesis of serious RS disease of early life must take into account the intrinsic pathogenic potential of the virus itself. Thus, in nursery outbreaks the virus caused pneumonia in children who lacked detectable serum antibody. If immunologic factors are important in lower respiratory tract disease they act to enhance the basic pathogenic effects of the virus. We propose that the interaction of serum antibody and viral antigens in the lungs plays an important role in producing the disease manifestations characteristic of RS illness of early infancy. If this is correct, then local respiratory tract antibody and serum antibody act in opposition, the former by preventing or moderating infection, while the latter probably participates in tissue damage.

Although the basis for immunological enhancement of RS virus disease of infancy is uncertain, several possibilities can probably be excluded. Type I (anaphylactic) and Type IV (cell-mediated delayed hypersensitivity) immune reactions are not involved, since neither reaginic antibody (IgE) nor cell-mediated sensitivity are transferred from mother to infant. For the same reason IgA and IgM antibodies could not participate. Maternally transmitted serum IgG antibody remains as the probable immunologic component of the disease complex: whether it reacts with virus antigen on the infected cell surface to produce a Type II immune (cytologic) reaction, or with soluble antigens of the virus in an extracellular location to effect a Type III immune reaction, is a matter for conjecture.

Immunological anomalies of respiratory syncytial virus infection of childhood

P. S. Gardner

Department of Virology, Royal Victoria Infirmary, Newcastle upon Tyne

Four unusual features occur in RSV infections of childhood:

(i) RSV infections are most frequent in the first year of life.
(ii) The severest infections are those in the first three months, often bronchiolitis and occasionally pneumonia.
(iii) RSV is frequently associated with deaths from respiratory infection in children under six months and also with 'cot deaths'.
(iv) These severe infections take place in the presence of maternal antibody.

It has been shown that children of three months can produce a humoral response to RSV but reinfection can occur, occasionally severely, in older children and even adults who have RSV antibody. The presence of RSV in the lungs of children who die of pneumonia associated with this virus is overwhelmingly shown by immunofluorescent examination, whereas by similar examination of bronchiolitis deaths or 'cot deaths' associated with RSV, only scanty localization is seen. In a recent series of 22 child deaths, RSV was isolated from 6 of 11 bronchiolitis, 3 of 5 'cot deaths' and 4 of 6 pneumonia, the other 2 also had a virus aetiology; 19 were less than six months old. One 'cot death' showed unexpected histological evidence of minimal bronchiolitic changes.

It is reported that children who received RSV vaccine developed severer illnesses than controls.

These apparently unrelated observations should have a common theme. The severity of the disease in infants could be explained by a hypersensitivity reaction, 'cot death' closely resembling acute anaphylaxis, bronchiolitis deaths and bronchiolitis itself being less acute reactions of the process. Possible mechanisms were investigated by immunofluorescent techniques and discussed in terms of Type I and Type III allergic reactions. Other factors to be taken into account are local antibody, the role of humoral antibody and previous infection with RSV.

Immunological aspects of pathogenesis

Résumé: presented by **S. Gard**

Prof. Gard opened the seminar by reviewing the immunological mechanisms likely to be involved in recovery from, and protection against, virus infections and those which might be involved in their pathogenesis. He concluded that the separation, by Coombs and Smith, of immunological reactions into beneficial and harmful ones was invalid and that any sort of reaction could be beneficial or harmful depending on circumstances. He referred particularly to the situation in measles where greater protection is conferred by gamma-globulin than by vaccination with inactivated vaccine, even though the latter yields higher levels of circulating neutralizing antibody. He discussed what the additional factor in gamma-globulin might be and suggested that it might be the 'serum protective factor' described in arbovirus infections by Price and his colleagues. This is reported to be thermolabile globulin-like material whose mode of action is not yet clear.

A good deal of discussion followed as to the nature and action of this factor, and attention was also drawn to the heterogeneity of antibodies and to work with pox viruses which suggested that virus released normally from an infected cell carried an additional antigen not present in other virus preparations. Antibody to this additional antigen appears to be essential to the protective action of antibody. Other possibilities considered were that the serum protective factor might be IgE or an antibody which reacted with parts of the virus surface not concerned with attachment to cells.

Dr. Perkins gave an account of the use of anti-thymocyte globulin to render mice susceptible to the implantation of HeLa cells. This was designed to provide a test for the oncogenicity of adenoviruses which might be used in vaccines. He had found that the antiserum not only made the mice highly susceptible to tumour formation but that antibody responses and protection following vaccination against tetanus were also greatly impaired. In the studies on tetanus toxin it is necessary to use anti-thymocyte globulin rather than serum because of some inhibitory factor in serum which appeared also to be present in normal rabbit serum.

In the discussion which followed it was pointed out that this system was likely to succeed as a test for oncogenicity in adenoviruses because it had previously been shown that thymectomized mice were a satisfactory test system. There was also a discussion on the possible nature of the inhibitory effect of serum in mouse-challenge tests with tetanus toxin.

Dr. Chanock described an extensive study of RSV infections, pointing out the high prevalence of disease in very young children with high levels of maternally derived antibody. Severe disease occurs only in this group, but increased severity of disease was also found in older children previously given an inactivated vaccine. Protection against infection depends on secreted IgA in the respiratory tract and this is not conferred maternally. He concluded that severe RSV disease is due to infection in the absence of secreted IgA but in the presence of high levels of circulating IgG and due to antigen-antibody interactions of types which could not yet be specified. He thought it unlikely that these interactions were with IgM or IgE. There was no evidence that complement was being consumed in these reactions.

Dr. Gardner followed with an account of studies of RSV infections in Newcastle. He divided the disease due to RSV into bronchiolitis and 'cot deaths' on one hand and pneumonia on the other. In autopsy material he had found less virus in the lungs and much less evidence of virus antigen by fluorescent antibody in the bronchiolitis and 'cot-death' cases than in pneumonia. Globulin was detected in cells in cases of bronchiolitis but not in pneumonia cases. He presented the idea that the bronchiolitis and 'cot deaths' might be due to second infections with RSV in which IgE reacted with the virus to cause allergic and even anaphylactic responses.

There was a great deal of discussion about the various possible immunological mechanisms which could account for the facts about the pathogenesis of RSV disease: evidence was against the suppression of cell-mediated immunity as an explanation—children with thymic aplasia survive beyond the age of RSV infections. Doubts were expressed as to whether bronchiolitis and pneumonia could be separated and as to whether bronchial spasm or constriction (rather than blockage) played a part in the disease. The design of vaccines was also discussed. Killed vaccines could no longer be used by parenteral injection and the evidence suggested that the local application of inactivated vaccines or of live vaccines made from 'cold' mutants might generate the secretory IgA necessary for protection.

Uni-directional flow ventilation in patient isolation

O. M. Lidwell and **A. G. Towers***

Central Public Health Laboratory, London, N.W.9

A bacteriologically clean atmosphere as part of the nursing environment has been an objective ever since the classical investigations of Pasteur and his successors demonstrated the ubiquity of microbial contamination. The ideal is more easily stated than attained and the inherent conditions of a hospital environment, the close physical juxtaposition of numbers of susceptible and hence at times infected individuals, increases the difficulties. The importance of airborne micro-organisms in the transmission of infection is always difficult to assess in a nursing situation involving repeated handling of patients with the many opportunities this affords for direct and indirect transfer of bacteria. The patient himself on arrival in the hospital or isolation unit is also colonized by or transiently carrying species of micro-organisms either internally or on the body surface or clothes. The relative importance of the different ways by which organisms responsible for clinical infection reach a susceptible locus must vary with the patient's condition, the environmental situation and the species of micro-organisms.

In this complex situation the provision of bacteriologically clean air for the patient may serve either or both of two purposes. It may benefit the patient by reducing the risk of infection and it may also provide an environment in which the transfer of micro-organisms by the other routes can be studied free from the complications of the otherwise ever-present possibilities of bacterial transfer due to airborne particles.

The approach towards a bacteria-free atmosphere implies excluding contaminated air from without, minimizing dispersal from potential sources of airborne micro-organisms which cannot be excluded from the enclosure, and removing as rapidly and completely as possible such contamination as it is disseminated.

Ventilation of a room with clean air does not usually displace an equivalent amount of the air previously present, together with any contamination it may carry. Turbulent movements associated with the inflow of the clean air, with air currents derived from thermal differences within the room or movement of objects and persons, lead to substantial intermixing. The extent of this is commonly such that the rate of removal

* Working on a Department of Health and Social Security Grant.

of airborne contamination approximates to a logarithmic process by which successive fractional reductions in the original level of contamination take equal times and complete elimination demands, in theory, an infinite time. In 1946, Bourdillon and Colebrook reported on the use of a mechanically ventilated room for dressing burns or major wounds. The principal features of this were (*a*) that the room was maintained at a slight positive pressure with respect to its surroundings so as to exclude air from outside and (*b*) that the air was introduced at ceiling level and was warmer than the room so as to 'form layers which are pushed downwards too slowly to cause a draught and which push the dirty air before them as they go'. In so far as this 'piston' effect is attained, complete removal of contaminated air can be achieved. Although the authors contemplated the use of quite high rates of air change in order to obtain the 'piston' effect, subsequent attempts to take advantage of this, for example in operating rooms, were not very successful owing to the air change being generally limited to 10–20 changes per hour on grounds of economy. The translational velocity of the 'piston', if attained under these conditions, is only of the order of 2–4 feet per minute, which is something like ten times smaller than the value needed to swamp the effects of common temperature differences and quiet movement (Lidwell *et al.* 1967). Industrial requirements for very clean atmospheres have in the last decade brought about widespread use of rooms employing the directed flow principle, commonly with a translational velocity either vertical or horizontal of about 100 feet per minute. The obliteration of large-scale turbulence in these systems has led to the use of the term 'laminar flow'. Small-scale turbulence, however, persists, both in the inlet air stream and in the wake of objects in the room. But so long as the residual turbulence is not large enough to cause significant air movement across the room, or even in an upstream direction, it does not impair the efficiency of the system in removing airborne contamination.

Over the same period there has been an intensification of interest in patient isolation and in the reduction of the risks of infection of particular groups of patients who are either especially prone to become infected or particularly at risk from the consequence of infection or both. The approach to this problem has involved the treatment of patients, especially with antibiotics, selection of nurses and staff to exclude carriers of particularly dreaded species or strains of micro-organisms, the extension of barrier-nursing procedures to minimize nurse–patient contact, the use of single rooms and special units, sometimes with elaborate ventilation provision to exclude extraneous airborne contamination, the nursing of patients in sealed plastic enclosures, by which means all contact with the outside environment is controlled.

The high level of air cleanliness possible with a uni-directional flow system of adequate velocity can be applied to the surgical situation (McDade *et al.* 1968; Charnley 1968) or to the nursing of an individual patient (Burke 1967; Michaelsen *et al.* 1968) but we have been interested in the possible advantages the system might have as a means of nursing several patients in the same room while maintaining a high degree of inter-patient isolation (*Laminar Flow Ventilation*, 1968).

A room arranged for several patients isolated from one another in respect of airborne contamination could have either horizontal or vertical air flow. The down-flow system has some possible advantages in respect of maintaining undisturbed flow lines, sedimentation follows the direction of the air flow and there is no contaminated downstream area arising from any source. The practical disadvantages are, however, considerable. Some additional building height is required and, unless one of the air ways can be sunk into the floor, the room must be entered up steps or a ramp. A perforated floor has many inconveniences and a further disadvantage is the extra air volume required. A single patient in an intensive nursing unit occupies about a 10-foot length of wall at the bed head. If this is, as is most natural for a horizontal air-flow system, the air inlet wall (see Figure 3) and the room is 8 foot high the volume of air required per patient, in cubic feet per minute, is 80 times the required linear velocity.

The floor area occupied by a patient in such a unit is likely to be at least 120–150 square feet, so that a vertical flow system will require 50–100 per cent more air than a horizontal system for the same velocity. Apart from the additional cost in providing the increased air volume, restricting the noise level to a volume acceptable both to patients and to the clinical staff is a major problem directly related to the air quantities required.

Cross transfer in a uni-directionally ventilated room

As there was little published data from which to make reliable deductions on the extent to which airborne contamination might be transported from one position to another across the direction of air flow, either as a result of the residual turbulence in the air stream or by disturbances of this arising from the movement of persons in the room, we studied this in a preliminary experiment (Lidwell and Towers 1969). By discharging large numbers (3×10^7 in a typical experiment) of particles carrying spores of *B. subtilis* var. *niger* we were able to detect the transfer of as little as one-tenth of the challenge on to a square foot of surface anywhere in the room. Since both economy and problems of noise call for as low an air velocity as practicable we examined the possibility of using air speeds substantially lower than the value commonly adopted in commercial installations, i.e. 100 linear feet per minute. Down to velocities as low as 35 feet per minute there was no detectable transfer of particles against or across the direction of air flow except when persons were moving about in the room, and even in these circumstances the numbers carried against the air flow were between 100 and 1000 times less than in a turbulently ventilated room (the equivalent ventilation rates in the room were between 130 and 360 per hour). The numbers carried across the air stream were about ten times greater, but transfer across the direction of air flow beyond the point reached by the persons moving about in the room did not extend to more than two or three feet beyond the limits of movement. Further reduction of the air velocity to 22 feet per minute, however, was associated with significantly greater transfer of particles in all circumstances and irregularities in the air flow due to thermal differences in the room were sometimes apparent.

Uni-directional ventilation in an Intensive Therapy Unit

The experimental room was then enlarged to accommodate two patients and re-erected in the intensive therapy ward of the thoracic surgical unit at Clare Hall Hospital, South Mimms, Hertfordshire. It was designed to operate at a linear velocity of 40 feet per minute and to meet a noise rating requirement of 35. There were some difficulties in meeting this last condition and the finally attained value reached the 45 NR level between 250 and 1000 Hz (Figure 1). This was accepted and during the first six months of operation there were no substantial complaints of the noise either by patients or by staff. This is, however, only marginally satisfactory and a further rise of five units in the NR associated with the intermittent use of extra humidifying equipment needed to maintain humidities about 50 per cent under all circumstances had led to complaints of noise. These noise levels appear to be much lower than those (about 80 db C) stated to be acceptable by Michaelson (1968).

In addition to the recirculating air, 250 cubic feet per minute or about six changes per hour of fresh air are introduced by a separate fan, which is duplicated in case of failure. Figure 2 shows the engineering arrangements in schematic form while the arrangement of the beds and the relationship to the rest of the ward is given in Figure 3.

This ward is used for the care of patients in the immediate post-operative period following major chest surgery and the normal period of stay is no more than three to five days.

Clinical evaluation of infection control measures is notoriously difficult, especially in specialized units with relatively few patients. The short-stay periods normal and desirable in intensive-care units increase these difficulties to the point of impracticability. We have therefore directed our efforts towards a bacteriological assessment, in the hope that we might be able to detect significant differences in the transfer and acquisition of micro-organisms between and by patients nursed in a uni-directionally ventilated environment compared with the experience of similar patients nursed in an open ward. We also hope to study the significance of particular nursing and other procedures in the transfer of micro-organisms from patient to patient and the effect of variations in these procedures.

Methods

Samples were taken for bacteriological examination from patients, staff and environment, including the air. In order to increase the numbers of traceable strains of micro-organisms, and since it is likely that the detailed routes of transfer differ for different species of micro-organisms, we have attempted typing differentiation of as many species as practicable.

Patient Specimens

Each patient admitted to the experimental room was examined daily, on the day before operation and as long as he or she remained in the room, except that in the case of the few patients whose stay extended over the weekend. Patients admitted to the room were operated on Monday or Tuesday and no specimens were taken on Saturday or Sunday. A similar number of patients nursed in the open ward after operation were examined in the same way. Initially no specimens were taken from the other patients in the open ward, but later nose swabs were taken from every patient at least once a week. The following specimens were obtained on each occasion.

1. A nasal swab using a swab moistened with nutrient broth.
2. The skin over the coccyx was sampled with a similar swab. As the majority of patients were nursed in an upright position this region could usually be examined easily.
3. The palmar surfaces of both thumbs and index fingers were imprinted on to the surface of serum agar plates containing phenolphthalein phosphate and on to blood agar plates.
4. 'Sweep' plates were obtained from the sheet over the abdomen and from the pillows. One edge of a plate of phenolphthalein phosphate serum agar was drawn across the fabric five times. Cotton bedding, including cellular cotton blankets, was used with cotton covers over the plastic-covered pillows.
5. Sputum specimens were examined regularly by the hospital laboratory as part of the normal control of the patient's condition.
6. Drainage tubes immediately after removal by the nursing staff were stirred in 10 ml. of transport medium which was stoppered and stored at room temperature until collected by an investigator.
7. Wounds were examined and sampled by the hospital laboratory.

Staff specimens

Nasal swabs, finger impressions and sweep plates from the belt area of the aprons and coats of the nursing staff were taken at weekly intervals and also at convenient times in relation to particular nursing activities.

Air samples

Petri dishes, $5\frac{1}{2}$ inches (14 cm.) in diameter or about $\frac{1}{6}$ square foot, containing serum phenolphthalein phosphate agar were exposed, for periods between one and six hours, in the experimental room and in the open ward.

In the experimental room slit samplers operating at 4 cubic feet (120 lites) per minute for periods up to two hours were used to obtain volumetric samples of the air downstream of each bed.

Surface samples

Swabs moistened with nutrient broth were rolled across defined areas (see Figure 3) of the walls, floor, door, bed tables and lockers in both the experimental room and in the wards. The curtains in the experimental room were sampled with sweep plates while in the drawn-up out-of-use position and swabs were also taken from the sink. Samples were taken at least weekly from each site.

Examination of specimens

All swabs, except from sinks, were plated immediately on blood agar and/or serum agar containing phenolphthalein phosphate. After incubation at 37°C the plates were read and colonies picked for identification and typing. Sink swabs were cultured in cetrimide broth for isolation of Pseudomonas.

Presumptive *Staphylococcus aureus* colonies on phenophthalein phosphate serum agar were tested for coagulase production. Positive strains were phage typed and tested for antibiotic resistance.

Presumptive colonies of Proteus, Klebsiella, Pseudomonas and a selection of colonies of white staphylococci were picked on to agar slopes for further examination as follows.

Proteus spp.

After plating on to McConkey medium a single colony was grown in broth and this was used for species identification, proticin typing and phage typing.

Klebsiella

These were characterized by a limited series of biochemical tests and then phage typed.

Pseudomonas

All cultures were typed serologically and a proportion were phage typed.

White staphylococci

A selection of strains were grouped biochemically, tested for antibiotic sensitivity and a further selection of staphylococcus group II and group V strains phage typed by Professor R. E. O. Williams at the Wright-Fleming Institute, Paddington, London.

Experiences in use

Between October 1968 and the end of February 1969 14 pairs of patients and two patients singly had been nursed in the experimental room, adding up to a total of 102 patient days. As stated earlier there were no substantial complaints of fan noise from the patients although the intermittent operation of the extra humidifier was undoubtedly disturbing. The medical staff also were not inconvenienced by the noise level, although this was certainly perceptible when listening to breath or heart sounds.

Owing to leaks in the system the air velocity actually attained during this period was only about 30 linear feet per minute. As part of the investigation of the air flow, recordings were made with a specially constructed anemometer with a rapid response time (about one second), which was able to differentiate between flow velocity in the vertical plane parallel to the direction of air flow and that at right angles to this, i.e. in a horizontal direction across the intended air flow such as could cause undesired transport of airborne particulate material between the beds. A short record from this instrument is shown in Figure 4. Random variation in the air velocity is apparent, but the magnitude of any transverse velocity is small and corresponds to a maximum angular deviation from the intended of no more than 16°.

Bacterial air samples

The air-flow pattern and particle transport in the room were also examined bacteriologically during vigorous mock nursing procedures on volunteer patients. The results of these tests are summarized in Table 1, from which it will be seen that transport of bacteria-carrying particles from the occupied bed due to these activities into the other bed area is at least an order of magnitude less than that observed when the directional flow was switched off and the room was ventilated in turbulent fashion by the fresh-air supply alone.

Checks in the unoccupied room after thorough cleaning by exposing settling plates for 12 hours and by volumetric air sampling showed values of less than 4×10^{-3} colonies/ft.^{2}/min. or less than 6×10^{-3} colonies per cubic foot. Since these values derived from only one colony for every two petri dishes used, most of this probably represents accidental contamination during handling.

Samples taken when the room was occupied by patients are analysed in Table 2. It is not easy to obtain representative air samples from a uni-directional flow system, since any liberated material largely remains in the flow line into which it was dispersed (v. Fox 1967). The values shown in the table may be compared with those found by Fox in mock operating-room experiments in a horizontally directed flow room at a velocity of 80 linear feet per minute, with those found in a vertical downflow operating room (McDade *et al.*, loc. cit.), and in conventionally ventilated operating rooms (e.g. Lidwell *et al.* 1967). The open ward at Clare Hall was fairly large, over 100,000 cubic feet, and was usually occupied by no more than four patients, who were generally quiet.

During the period described here no patient in the experimental room was dispersing appreciable numbers of an identifiable strain of bacteria, so that we have so far no records (apart from a single colony of *Staph. aureus* isolated directly downstream of a patient who was a carrier of the strain) from which we can confirm the absence of air transport from one bed to the other under normal working conditions.

Nursing activity

These patients needed considerable nursing attention. It is not easy to provide a quantitative index of this, but in order to obtain some indication of the amount of activity magnetically operated door switches were fitted and coupled to a system which recorded both the number of times the door was opened and the aggregate duration of door opening. Over sixteen 24-hour periods during which five pairs of patients were being nursed in the room the number of door openings in a 24-hour period ranged from 200 to 600 with an average of 360 and the doors were open for an average of 48 minutes or about eight seconds for each opening. This corresponds to 15 entries or exits per hour and is some reflection of the time load imposed by nursing patients in separate rooms.

Patient bacteriology

Over the period 12 Oct. 1968 to 24 Aug. 1969 observations were made on a total of 101 patients, 23 pairs and five single patients nursed in the experimental room and 50 patients nursed in the open part of the ward. The average number of days of observation for each patient post-operatively was 2·85. A summary of the bacterial isolation from these patients is given in Tables 3 and 4.

Pre-operative nasal carriage of *Staphylococcus aureus* was unexpectedly low among the patients nursed in the open part of the ward. No carriers were found among the last 25 of these patients examined. There seemed to be no reason for this and it was probably a chance occurrence. The overall deficiency was confined to carriage of strains sensitive to all antibiotics and of those resistant to penicillins only. There were no substantial changes in the frequency of nasal carriage of coagulase-positive staphylococci post-operatively but recoveries from the fingers were markedly more frequent. There were also a substantial number of isolations of this organism from the bed linen, even before the bed had been occupied by the patient.

Pre-operative isolation of the three gram-negative species studied was relatively uncommon and never found on the unoccupied bed. Post-operatively these organisms were isolated much more frequently. *Proteus mirabilis* was the species most often recovered, especially in the experimental room. Klebsiellas were nearly as often found in the open ward together with a few Pseudomonads. Whether the different distribution of the gram-negative species in the two locations has any significance is not easily answered from the relatively small number of isolations made. The principal difference between the two environments was that the relative humidity in the experimental room was maintained constant at near 60 per cent while in the open ward was more often around 40 per cent during the autumn, winter and spring months.

The study of the apparent acquisition of bacteria by the patients is more interesting. For the purpose of this investigation an apparent acquisition is defined as the recovery from a patient specimen of a species or strain of bacterium not found in the pre-operative samples from the same patient. In view of the comparatively inefficient nature of the sampling techniques when only small numbers of micro-organisms are involved it is clear that a proportion of these 'apparent acquisitions' must be spurious and result from a failure to detect carriage of the organisms at a low level pre-operatively. We have not, however, at present any way of estimating the frequency of this. The numbers of apparent acquisitions are shown in Table 5 according to the species of bacterium isolated, the site of first isolation and the frequency with which the strain was isolated from the individual patients. It will be seen that most of the 'acquisitions' refer to a single isolation from a single site on the patient. It would seem unlikely that many of these indicate colonization of the patient. Where recoveries of the same strain were made on several occasions or from several sites, however, actual colonization may have taken place. Only one acquisition was associated with sepsis, when the wound of a patient in the experimental room became colonized with a coagulase-positive staphylococcus around a drain on the third post-operative day. The strain had been carried by other patients and members of staff, but no carrier was detected during the period immediately prior to colonization.

Acquisitions were rather more numerous both with staphylococcal strains and with the gram-negative species among the patients nursed in the open ward than among those nursed in the experimental room, although the overall difference 38:28 is not significant. Examination of the rates of acquisition over the post-operative period, Table 6, shows that this difference was most pronounced, especially for the staphylococcal strain, for acquisitions occurring two or more days post-operatively. Since acquisitions on the first post-operative day perhaps mainly reflect handling concerned with the operation itself and include an undue proportion of 'spurious acquisitions' of the kind referred to above, this might suggest a bigger difference between the risk of acquisition in the two environments. The actual number of acquisitions in the two locations after the first post-operative day were 27 and 13 respectively.

For acquisition of staphylococci the difference is concentrated in the multi-resistant strains which are characteristically dispersed in the hospital environment by patient carriers, although their frequent isolation from bed linen suggests contact transfer via the staff, perhaps on their clothes (Speers *et al.* 1969). Strains of proteus were the most common gram-negative organisms acquired in the experimental room (75 per cent) but Klebsiellas accounted for half the acquisitions in the open ward. As with the overall number of isolations the significance, if any, of the different distribution of gram-negative species acquired in the two locations is uncertain. While two-thirds of the staphylococcal acquisitions represented a single isolation, over half the acquisitions of gram-negative strains involved multiple isolations, usually on more than one day.

Just over half of the acquisitions of staphylococcal strains, both by patients in the open ward and by those nursed in the experimental room, could be related to staff carriage of similar strains at the relevant time. In no instance could other patients be implicated in this way and the sources of the remaining acquisitions remain unknown. It must be remembered that both in the ward and in the rest of the hospital there were other unexamined patients and that there was continuous interchange of both staff and patients between the intensive-therapy ward and the other wards of the hospital.

No plausible source could be traced for the great majority of acquisitions of gram-negative strains. In a few instances staff carriers may have been involved. Only one of the few possible patient sources represented a patient actually being nursed in the same room as the recipient at the relevant time. We have not been able to examine faecal specimens, but isolations from the skin over the coccyx, which we hoped might serve as an indication of dispersal from the anal region, were not numerous.

Discussion

The observations described above refer to only a small number of patients examined over a relatively short period of time and it would be rash to assume that they are representative. During the time in question patients in the experimental room were nursed in exactly the same way as those in the open ward. No additional precautions were taken to reduce the risks of bacterial transfer during nursing procedures. This was done in order to see the extent to which it was possible to find evidence of bacterial contamination and colonization of patients nursed to a conventional routine. Evidence for the value of any variation in procedure would be obtainable only if a measurable amount of acquisition was demonstrable when standard methods were used. The figures given in Tables 5 and 6 show definite evidence of bacterial acquisition, mostly of a transient character. Successful development of typing systems for the white staphylococci, the most frequently isolated species, would almost certainly multiply these instances. The problem here is not simply that of classifying individual strains but also of separating the population into sub-groups, identifiable in the presence of each other, such that an individual is unlikely to carry more than one strain of a given sub-group at any one time. This is a generally necessary condition for any practical possibility of relating acquisition to sources. Although a considerable number of apparent acquisitions were recorded it has not been possible to identify possible sources for the great majority. It is not clear how far this may be due to failure to isolate from the pre-operative samples a strain already carried by the patient on one or other site, examined or unexamined. The distribution of the frequencies with which a strain once isolated was recovered on one or more subsequent occasions would appear to make this an unlikely cause for more than a fraction of instances. The trauma associated with operations may, however, lead to a change in the distribution of micro-organisms over the body surfaces. Within the hospital as a whole there were clearly many more possible carriers than those actually examined who comprised the actual ward staff, i.e. those in close contact with the patients in question. We are extending the range of staff examinations, but the manageable bacteriological load is limited, especially in regard to some of the typing procedures.

The frequent changes in nursing staff and the short stay of the patients in the ward with its consequent rapid turnover made it difficult to maintain full coverage of the possible sources of bacterial strains.

Apparent acquisition of gram negative species, especially Proteus strains and Klebsiellas, was unexpectedly frequent. It is not clear whether there is any significance in the more frequent appearance of Proteus strains in the experimental room and Klebsiellas in the open ward.

That no satisfactory approach to the problems of assessing isolation facilities and procedures has yet been evolved, in spite of the considerable number of published studies, is an indication that no single solution is likely to be found. The diversified bacteriological approach which we are attempting to develop tries to counter the relatively small amount of time spent by individual patients in this specialized unit and the random time distribution of bacterial strains in an hospital environment by multiplying the number of tracer organisms employed. Since the relative importance of the different routes of transfer almost certainly differs for different bacterial species, this method might also avoid some of the risks of misleading generalization from observations on a single species. The extent to which this approach can be carried through depends both on the technical possibilities of isolation and strain discrimination of the available species and on the laboratory resources which can be committed to a given investigation. An alternative approach especially for direct clinical assessment, which is the ultimate criterion, involves co-operative standardization of installations, procedures and recording among a large number of units. This is a formidable requirement, but some European moves in this direction are under way.

As far as the development of uni-directional air-flow isolation systems are concerned, it is clear that if they afford any real clinical advantage then there are no practical obstacles other than financial ones. In particular the facility of nursing more than one patient in air isolation in the same room can be extended to any desired extent with all the advantages of easy nursing access which this implies. The problem will remain, however, of reconciling easy access in a busy environment with the maintenance of what are probably the more important isolation precautions, namely those which limit contact transfer of micro-organisms by handling.

ACKNOWLEDGEMENT

We would like to thank the medical, nursing, and engineering staff of the hospital for their help in this investigation.

SUMMARY

It should be possible to nurse several patients together within a single room and at the same time to isolate them effectively from each other with regard to airborne contamination by ventilating the room with a uni-directional flow system.

Bacteriological studies have been carried out on 51 patients who had undergone major thoracic surgery and who were nursed during their first few post-operative days in a two-bedded room of this kind furnished with a horizontally directed uni-directional flow system designed for an air velocity of 40 feet per minute. Simultaneous observations were made on 50 patients who had similar operations but were nursed in an open ward. There were no practical difficulties, but more extended studies will be necessary before any conclusions can be drawn as to the importance of different routes of patient infection and whether the system has any advantages over alternative methods of isolation.

REFERENCES

Bourdillon, R. B. and Colebrook, L. (1946). Air hygiene in dressing rooms for burns or major wounds. *Lancet*, **i,** 601.

Burke, J. F. (1967). A new approach to isolation. *Hospital Practice*, **2,** No. 2, Feb.

Charnley, J. (1968). *Hospital Management*, Modern British Operating Theatres. *Supplement*, September, p. 44.

Fox, D. G. (1967). An empirical study of the application of a horizontal unidirectional air flow system for a hospital operating room. Doctoral thesis submitted to the University of Minnesota, Minneapolis, U.S.A.

Laminar Flow Ventilation (1968). Laminar flow ventilation in an intensive therapy unit at South Mimms, Hertfordshire. *Nursing Times* (London), 27/9/68, p. 1317.

Lidwell, O. M., Richards, I. D. G. and Polakoff, S. (1967). Comparison of three ventilating systems in an operating room. *J. Hyg. Camb.*, **65,** 193.

Lidwell, O. M. and Towers, A. G. (1969). Protection from microbial contamination in a room ventilated by a unidirectional air flow. *J. Hyg., Camb.*, **67,** 95.

McDade, J. J., Whitcomb, J. G., Ryptia, E. W., Whitfield, W. J. and Franklin, C. M. (1968). Microbiological studies in a vertical laminar air flow surgery. *J.A.M.A.*, **203,** 125.

Michaelsen, G. S., Halbert, M. M., Sorenson, S. D. and Vesley, D. (1968). Development of an open isolation system for the care of low resistance hospital patients. *Report, School of Public Health*, University of Minnesota, Minneapolis, U.S.A.

National Physical Laboratory (1962). *The Control of Noise*, H.M.S.O., London.

Speers, R., Shooter, R. A., Gaya, H., Patel, N. and Hewitt, J. H. (1969). Contamination of nurses' uniforms with *Staphylococcus aureus*. *Lancet*, **ii,** 233.

TABLE 1

Dispersion by volunteers simulating patients (relative bacterial recoveries)

Sampling position	*Uni-directional air flow (30 feet per minute)*	*Turbulent ventilation (six air changes per hour)*
Close to patient bed	100	100
Directly downstream of patient bed	26	22
Corner downstream proximal to patient bed	11	15
Centre line of room by downstream wall	1·2	19
Centre line of room by input filters	0·2	58
Close to or downstream of unoccupied bed	0·9 0·8[a]	12

[a] Refers to samples taken with the slit samplers (see text); all other figures relate to sampling by settling.

TABLE 2

Contamination levels in the experimental room and other places

Place		Sampling position		
		By input filters	*Near patient bed or operation area*	*Downstream*
Experimental room		0·05	2·0	1·4
Open ward		—	7·2	—
Operating room				
1. Horizontal uni-directional airflow	(*a*)	—	*0·21*	*0·7*
(mock operations)	(*b*)	—	*0·06*	*1·0*
	(*c*)	—	*0·02*	*0·8*
2. Vertical uni-directional downflow	(*d*)	—	*0·6*	—
3. Conventionally ventilated	(*a*)	—	*5*	—
	(*b*)	—	*0·6*	—
	(*c*)	—	*0·7*	—
	(*d*)	—	*2·4*	—
	(*e*)	—	5·8, 3·6	—

Figures are given as colony-forming particles settling per square foot of exposed surface per minute or, in italics, collected per cubic foot of air sampled. The range of particle size of the airborne bacteria carrying particles leads to similar values for these two quantities when parallel samples are taken.

(*a*), (*b*) and (*c*) figures from Fox (1967)—(*a*) personnel in street clothes, (*b*) in scrub suits with cap, mask and gloves, (*c*) full surgical dress; (*d*) figures from McDade *et al.* (1968); (*e*) Lidwell, Richards and Polakoff (1967).

TABLE 3

Percentage positive isolations, Staphylococcus aureus

Location	Site				Strain				Patient days sampled
	N	*F*	*Co*	*Li*	*S*	*P*	*R*	*Any*	
Pre-op.									
Open ward	10	2	0	4[a]	2	4	6	12	46
UDF room	24	0	4	6[b]	12	12	4	28	49
Post-op.									
Open ward	11	7	0	15	4	8	11	23	140
UDF room	20	6	1	8	14	9	4	27	148

[a] Bedclothes sampled before patient was put into bed.

N = nose, F = fingers, Co = skin over the coccyx, Li = bed linen. The staphylococcal strains were differentiated according to their sensitivity to antibiotics. S = sensitive to all antibiotics, P = resistant to penicillins that are penicillinase sensitive, R = resistant to any antibiotic other than the above penicillins, usually resistant to a range of antibiotics, UDF = experimental room ventilated with a uni-directional air flow.

TABLE 4

Percentage positive isolations, gram negatives

	Site				Species				
Location	*N*	*F*	*Co*	*Li*	*Pr*	*Kl*	*Ps*	*Any*	*Patient days sampled*
Pre-op.									
Open ward	8	0	0	0[a]	8	0	0	8	46
UDF room	2	0	4	0[a]	4	2	0	6	49
Post-op.									
Open ward	11	6	2	8	11	8	4	23	140
UDF room	8	7	4	5	13	3	0	16	148

[a] Bedclothes sampled before patient was put into bed.
Sites and locations as described for Table 3.
Pr = Proteus spp., Kl = Klebsiella, Ps = *Pseudomonas aeruginosa.*

TABLE 5

Number of apparent acquisitions of bacteria

Location and species		1st site of isolation					Frequency of isolation			
		N	*F*	*Co*	*Li*	*Any*	*S*	*SD*	*SS*	*M*
Open ward										
Staphylococcus aureus	S	1	1	0	1	3	3	0	0	0
	P	$2\frac{1}{2}$	$1\frac{1}{2}$	0	3	7	5	1	1	0
	R	1	$\frac{1}{2}$	$\frac{1}{2}$	6	8	4	2	0	2
Any *Staphylococcus aureus*		$4\frac{1}{2}$	3	$\frac{1}{2}$	10	18	12	3	1	2
Proteus		5	0	0	2	7	2	1	3	1
Klebsiella		$1\frac{1}{2}$	2	$2\frac{1}{2}$	4	10	5	1	0	4
Pseudomonas		1	$\frac{1}{2}$	0	$1\frac{1}{2}$	3	1	1	1	0
Any gram negative		$7\frac{1}{2}$	$2\frac{1}{2}$	$2\frac{1}{2}$	$7\frac{1}{2}$	20	8	3	4	5
UDF room										
Staphylococcus aureus	S	2	0	0	1	3	2	0	1	0
	P	$1\frac{1}{2}$	$2\frac{1}{2}$	0	3	7	5	1	0	1
	R	$\frac{1}{3}$	$\frac{1}{3}$	0	$1\frac{1}{3}$	2	1	0	0	1
Any *Staphylococcus aureus*		$3\frac{5}{6}$	$2\frac{5}{6}$	0	$5\frac{1}{3}$	12	8	1	1	2
Proteus		5	$2\frac{1}{2}$	2	$2\frac{1}{2}$	12	4	1	2	5
Klebsiella		$\frac{1}{2}$	$2\frac{1}{2}$	1	0	4	3	1	0	0
Pseudomonas		0	0	0	0	0	0	0	0	0
Any gram negative		$5\frac{1}{2}$	5	3	$2\frac{1}{2}$	16	7	2	0	5

Sites and strains of *Staphylococcus aureus* as described for Table 3.

When the strain was first isolated from more than one site at the same time this has been recorded as the appropriate fraction under the 1st site of isolation.

UDF = experimental room ventilated with a uni-directional air flow.

S = single isolation only.

SD = isolated from more than one site but on one day only.

SS = isolated on more than one day but from one site only.

M = isolated from more than one site and on more than one day.

TABLE 6

Rates of apparent acquisition of bacteria (/100 days)

Location and species		Day of 1st isolation: 1st post-operative	2nd	3rd–5th	Any	Possible source (number of acquisitions): Staff	Other patients	Unknown
Open ward								
Staphylococcus aureus	S	2	2	2	2	1	0	2
	P	4	2	8	5	4	0	3
	R	4	6	6	6	5	0	3
Any *Staphylococcus aureus*		10	10	16	13	10	0	8
Proteus		4	6	4	5	3	1	3
Klebsiella		6	2	12	7	0	0	10
Pseudomonas		2	2	2	2	0	0	3
Any gram negative		12	10	18	14	3	1	16
UDF room								
Staphylococcus aureus	S	2	0	4	2	2	0	1
	P	8	4	2	5	4	0	3
	R	4	0	0	1·5	0	0	2
Any *Staphylococcus aureus*		14	4	6	8	6	0	6
Proteus		12	8	4	8	2	4	6
Klebsiella		4	2	2	3	0	0	4
Pseudomonas		0	0	0	0	0	0	0
Any gram negative		16	10	6	11	2	4	10

Sites, locations and strains of *Staphylococcus aureus* as described for Table 3.
Other patients have been listed as possible sources when the strain had been previously recovered from a patient; on only one occasion was the possible source so defined, present at the same time and in the same room as the recipient.
The rates given in the left-hand columns of the table have been rounded to the nearest half-unit.

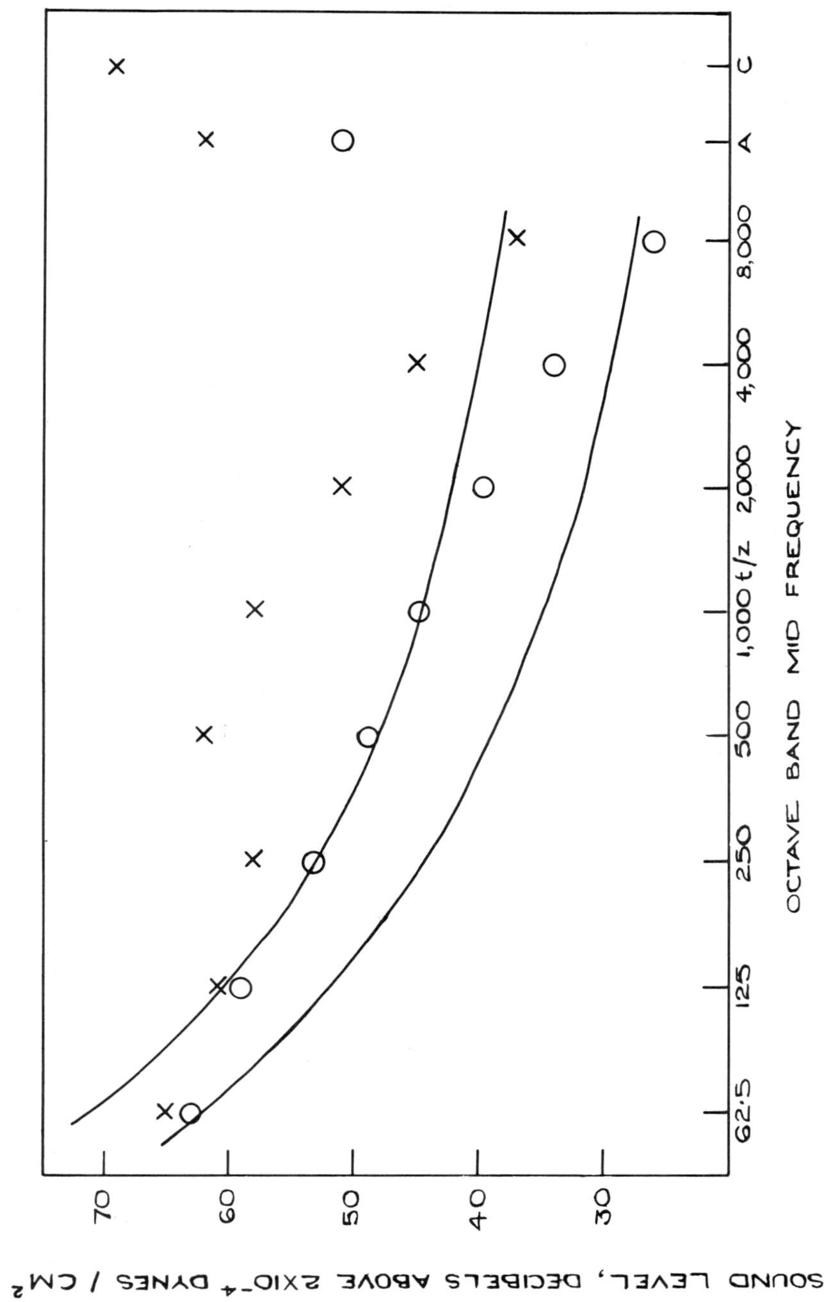
70
60
50
40
30
SOUND LEVEL, DECIBELS ABOVE 2X10-4 DYNES / CM2
62·5
125
250
500
1,000 t/z
2,000
4,000
8,000
A
C
OCTAVE BAND MID FREQUENCY

Figure 1

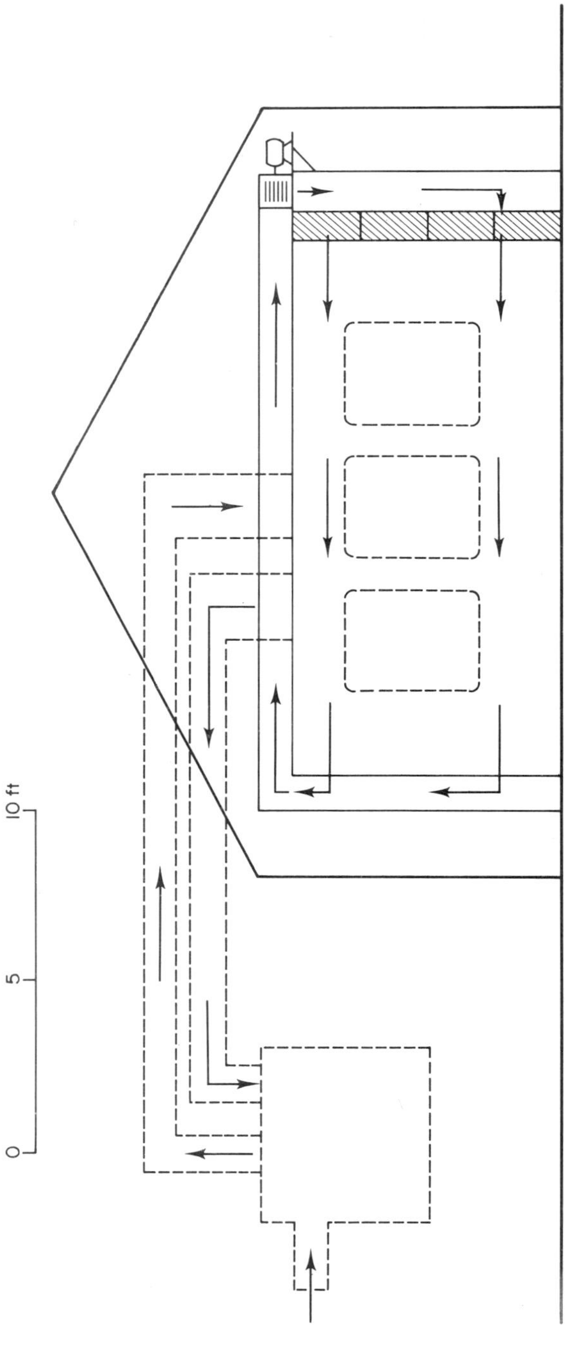

Figure 2

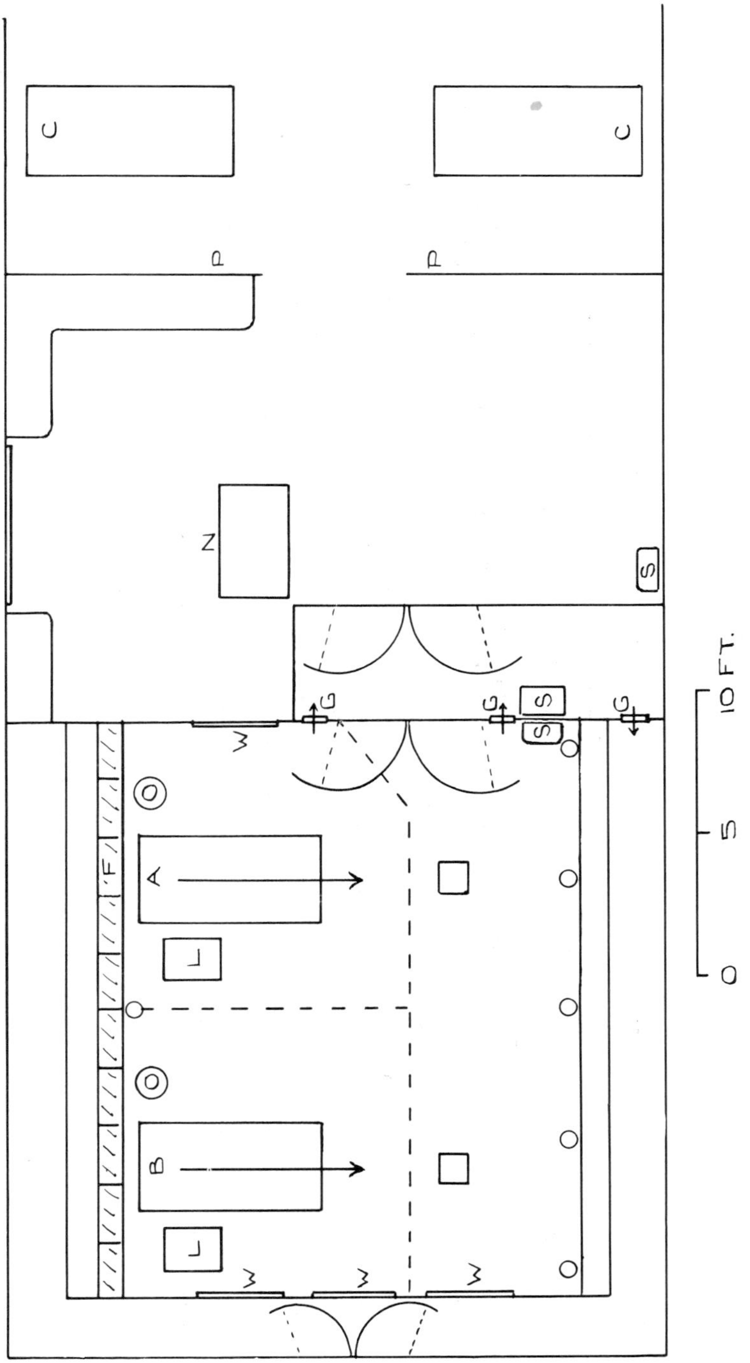

Figure 3

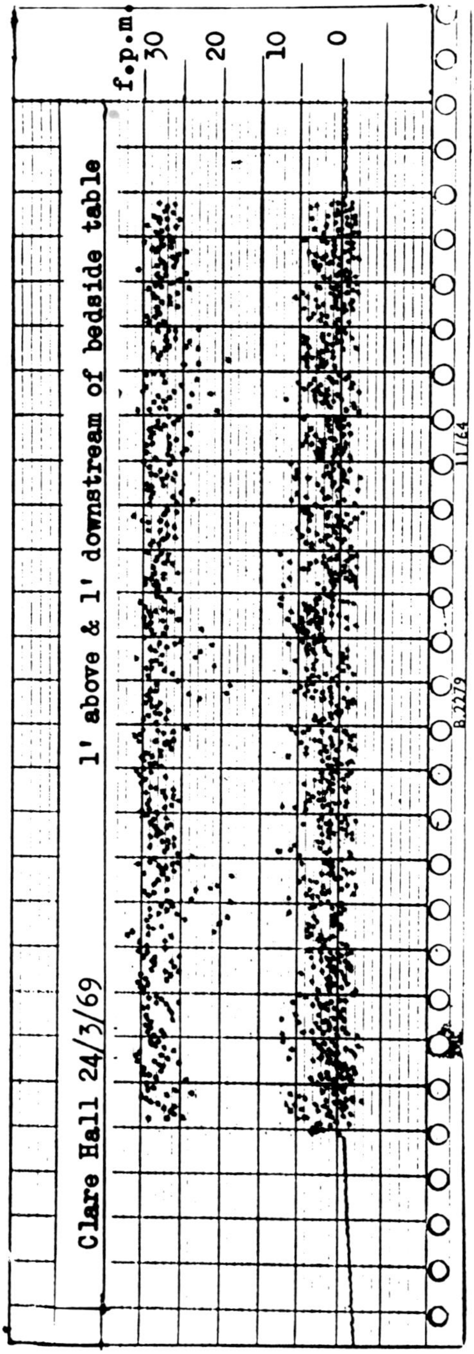
Clare Hall 24/3/69
1' above & 1' downstream of bedside table
f.p.m.
30
20
10
0

Figure 4

Controlled trials of patient-isolation in a burns unit

E. J. L. Lowbury

M.R.C. Industrial Injuries and Burns Research Unit, Birmingham Accident Hospital

---

Isolation is generally regarded as an important measure not only for the segregation of infected patients in hospital but also for the protection of those who are highly susceptible to infection. In a recent survey (Report, 1965) it was estimated that such 'protective isolation' was required for 8·1 per cent of patients admitted to general surgical wards. All patients having immunosuppressive drugs for organ transplantation or cancer chemotherapy fall into this category; other susceptible patients for whom protective isolation has been thought desirable include those with extensive burns (e.g. Colebrook 1950).

The usual method of providing isolation is in single-bedded isolation rooms, preferably with positive pressure mechanical ventilation to keep out airborne bacteria; this method has been developed for use in 'ultra-clean' wards (e.g. Bagshawe 1964; James *et al.* 1967; Woodruff *et al.* 1968). Another method is the use of plastic ventilated isolators, in which techniques developed for the study of germ-free animals have been adapted for the isolation of patients (Levenson *et al.* 1964; Haynes and Hench 1966; Cason *et al.* 1966; Robertson *et al.* 1968).

Little evidence has so far been obtained on the efficacy of protective isolation. Bacterial counts of air in mechanically ventilated rooms are shown to be lower than they are in rooms without mechanical ventilation (see Figure 1), but the effects of such reduction of bacterial contaminants on sepsis rates are more difficult to demonstrate. In a recently opened, sub-divided ward for thoracic surgery, Lidwell and others (1966) found a low incidence of wound infection, an absence of nasal acquisition of resistant staphylococci and fewer airborne staphylococci than had been found in open wards previously used. It is uncertain, however, which factors are responsible for the reduction in sepsis in environments where a number of changes in practice as well as in structure have occurred.

Earlier studies in the Burns Unit at Birmingham showed a significantly lower incidence of infection in patients whose dressings were changed in a plenum-ventilated room than in a comparable series of patients dressed in the same room with no mechanical ventilation (Lowbury 1954). On the other hand, subdivision of an open burns ward into one with eight two-bedded rooms and two single-bedded rooms did not lead to any reduction in infection with *Staph. aureus* and *Ps. aeruginosa*; the subsequent installation of partial recirculation air-conditioners into half of these isolation rooms caused approximately 50 per cent reduction in airborne *Staph. aureus* in these rooms, but this was not associated with a lower infection rate in patients in the air-conditioned rooms (Cason *et al.* 1966).

It was inferred from these findings that protection of patients in a burns ward required much more intensive isolation methods than those which had been studied. Patient isolators seemed to offer the best prospects of physical protection against both airborne and contact transfer of bacteria. A trial was therefore started in order to assess the efficacy of this method of treatment and its practicability for patients with burns.

MATERIALS AND METHODS

Isolators

After preliminary studies with other equipment, the Vickers patient bed isolation system (Model 55) was used. This consists of a 'canopy' of transparent, flexible polyvinyl chloride which completely encloses the patient. The canopy is suspended on a metal framework over the bed. It is inflated and ventilated by air supplied through flexible ducts from a quiet centrifugal fan unit; the air passes through a coarse spun nylon pre-filter and then through a main glass-paper filter estimated to have an efficiency of more than 99·9 per cent against particles with a diameter down to 0·58 μ. The canopy has, on each side, a row of five glove ports for handling the patient and other manipulations and an air-lock transfer port with inner and outer zip fasteners, one pouch for supply of clean and one for removal of dirty materials. There is a zip fastener along the full length of the canopy to admit the patient.

During the study many improvements in the design of the isolator have been suggested and adopted on the basis of experience by nurses in using it; a number of difficulties have been overcome by these modifications. Figures 2 and 3 show a recent model with a half-jacket and ventilated helmet to facilitate nursing and clinical management.

Bacteriological study of isolators

Settle-plates of phenolphthalein diphosphate agar (Barber and Kuper 1951) were exposed for six hours in empty isolators with coarse pre-filter and filter, with the pre-filter only, and with no filter in the air supply line; settle-plates were exposed at the same time in the open ward near the isolators.

Bacterial counts of air taken with a large slit sampler through flexible tubing from the interior of the isolator gave misleading results, apparently because of negative pressure which led to the entry of unfiltered air. An Andersen sampler was therefore used. Counts of total bacteria and presumptive *Staph. aureus* on 30-minute and 60-minute samplings inside and outside the isolators with and without filters were obtained; the size range of bacteria-carrying particles in the samples from each source was estimated.

Controlled trials

Patients with flesh burns of between 5 per cent and 30 per cent of the body surface were allocated in rotation, when eligible on other grounds, to three treatment groups: (1) for treatment in an isolator with pre-filter and main filter; (2) for treatment in an isolator from which, in the first phase of the study (one year), the main filter was removed, and in the second phase (one year) both pre-filter and main filter were removed; and (3) for treatment in the same ward but not in an isolator (control group). Patients were in the isolator groups and the control groups for periods up to three weeks. The group of patients treated in isolators without filters was included to provide information about the relative importance of airborne and contact infection, on the assumption that the isolator without filters contained numbers of bacteria similar to those in the air of the open ward; the group in isolators equipped with a coarse pre-filter only was included to provide evidence on the degree of filtration needed for protection against airborne bacteria.

Local treatment of burns was by the exposure method or (in most cases) by the application of dressings with penicillin cream (1000 units per G) to prevent infection with *Streptococcus pyogenes*; cloxacillin (250 mg. six-hourly) was given by mouth to all patients during the first week (partly as prophylaxis against tetanus). Specific chemoprophylaxis against other organisms was not used; patients in whom it seemed important to use such measures were not admitted to the trial of isolators.

Bacteriological swabs were taken from burns if they were exposed, and otherwise at change of dressings; the swabs were examined by methods described elsewhere (Lowbury 1960; Gason *et al.* 1966). Nasal swabs were taken daily and examined for *Staph. aureus*, which were tested for sensitivity to a range of antibiotics. Rectal swabs were taken from all patients on admission and examined for gram-negative bacilli and *Staph. aureus*.

RESULTS

An interim report on progress in this study is presented here.

Bacteriology of air in isolators.

Table 1 shows the mean settle-plate counts in the open ward and in isolators with and without filters. The settle-plate counts in the isolator with filters were very low, but those in the isolators without filters were also much lower than those obtained in the open ward.

The mean bacterial counts per cubic foot of air taken with an Anderson sampler are shown in Table 2; the counts from inside the isolator without filters were lower (in one experiment much lower) than those from the outside air, but the difference was less than it was on settle-plates. There was little difference in the counts from the isolator with a coarse pre-filter only and from the isolator with no filter.

In Table 3 the distribution of sizes of bacteria-carrying particles is presented. This shows that most of the larger particles did not reach the inside of the isolator without filters; the surprisingly low settle-plate counts inside the isolator without filters are consistent with the lower mean size of bacteria-carrying particles in the air of this isolator compared with that of bacteria-carrying particles in the open ward air.

Controlled trial

Table 4 shows the numbers of patients in the first phase (one year) and in the beginning of the second phase of the trial from whose burns *Ps. aeruginosa* was isolated. The proportion of patients treated in isolators who became infected with *Ps. aeruginosa* was significantly smaller than that of patients in the other groups. The proportions of patients in the two isolator groups who acquired *Ps. aeruginosa* in their burns were similar (though slightly higher in isolators with pre-filters only). From the results in the first phase of the trial it appeared that the presence of a fine filter contributed little, if at all, to the effectiveness of the isolator in preventing infection with *Ps. aeruginosa*; few patients have been admitted to the trial in its second phase, and no inference can yet be made about the use of isolators with unfiltered air.

Patients who acquire multiple-resistant *Staph. aureus* were slightly, but not significantly, more numerous in the control than in the isolator groups, especially during the first week. A large proportion of the patients acquired resistant staphylococci while treated in isolators, especially in the group supplied with filtered air (see Table 5). Nasal acquisition of *Staph. aureus* showed a similar pattern, with a slightly but not significantly lower incidence and a later acquisition of multiple-resistant *Staph. aureus* in patients treated in isolators than in those treated in the open ward (Table 6).

The comparability of patients in different groups of the trial is shown in Table 7.

DISCUSSION

Preliminary assessments in the trial described above showed that isolators had some value in protecting burns against *Ps. aeruginosa*, but were much less effective in keeping the burns and nares of patients free from hospital strains of *Staph. aureus*. The presence of a high-performance filter made little if any difference to the results; the effect of removing both filters on infection rates is at present under investigation. Tests with settle-plates showed very low counts in the isolator without filters—a finding that was associated with a smaller mean bacteria-carrying particle size in the air of this isolator than in open ward air. From this we infer that patients in isolators without filters are not exposed to the same hazard of airborne contamination as that to which patients in the open ward are exposed, and the relative importance of airborne and contact transfer can therefore not be satisfactorily assessed from the trials of isolators. To throw light on this problem a bed protected against airborne contamination by 'air curtains' will be studied in a third phase of the trials; patients in this bed will be exposed to the same hazards of contact contamination as those in ordinary beds, but should have some protection against the airborne transfer of bacteria.

The use of local chemoprophylactic agents active against *Ps. aeruginosa* and *Staph. aureus* has been avoided in these trials, so that the role of isolators can be independently assessed; extensive burns in which the exclusion of local chemoprophylactic could not be justified were therefore not included in the trial, but these are the burns for which the use of isolators is most likely to be reserved if the method is found efficacious. Although the isolator simplified aseptic method in some ways (e.g. by making it unecessary for the nurse to put on a gown and mask), there are many difficulties and complications in their use, and these would be still greater in the management of more extensively burned patients, including infants and large adults, than in the children and small adults with burns of moderate extent who were studied in our series.

Undoubtedly the use of isolators can be made much more effective than it was in our study by the exclusion of any unsterilized materials from the isolator. Such isolation is rational for the protection of patients receiving immunosuppressive treatment, but it is rarely justified in the treatment of patients with burns, for whom other effective methods of prophylaxis are available and in whom loss of antibacterial resistance is rarely so acute and in most cases of relatively short duration. It may, however, be useful to have this method of protection available for the treatment of patients with burns in whom other methods of prophylaxis are, for some reason, unsuitable; its effectiveness cannot be lost through the emergence of bacteria resistant to antibiotics and other chemical agents, and most patients tolerate short periods of enclosure in an isolator very well.

SUMMARY

A controlled trial in progress on patients with burns has shown that a significantly smaller proportion became infected with *Ps. aeruginosa* when they were treated in plastic ventilated bed-isolators than when their beds were in the open ward; this effect was found in isolators with coarse pre-filters only as well as in isolators protected against airborne contamination by coarse and fine filters in the air supply. Isolators without filters are under study.

Infection with *Staph. aureus* resistant to two or more antibiotics (hospital strains) was often found in patients treated in isolators; though this occurred more often in patients of the control series, who were not in isolators, the difference was not significant.

Air samples showed lower counts of bacteria, lower mean bacteria-carrying particle-sizes and much lower settle-plate counts inside an isolator without filters than in open ward air.

Prevention of manual contamination by the use of glove ports (and, possibly, of airborne contamination by provision of filtered air) helped to protect patients against *Ps. aeruginosa.* To achieve protection also against *Staph. aureus* sterile supplies of food and other materials to patients in the isolators would be required. The practicability of such a germ-free isolator for some patients with burns is discussed.

ACKNOWLEDGEMENTS

I am grateful to Vickers Ltd. for the supply of isolators during our pilot studies, to the Department of Health and Social Security for their support of this research, to Dr. P. C. Trexler for valuable suggestions, to Miss P. M. Ford, S.R.N., for studies on the nursing of patients in the isolators, to Mr J. R. Babb, F.I.M.L.T., for assistance with laboratory studies, to Mr W. H. Cater for technical assistance with the isolators, and to the nursing and clinical staff for their co-operation.

REFERENCES

Bagshawe, K. D. (1964). Ultra clean ward for cancer chemotherapy, *Brit. Med. J.,* **ii,** 871.

Barber, M. and Kuper, S. W. A. (1951). Identification of *Staph. pyogenes* by the phosphatase reaction, *J. Path. Bact.,* **63,** 65.

Cason, J. S., Jackson, D. M., Lowbury, E. J. L and Ricketts, C. R. (1966). Antiseptic and aseptic prophylaxis for burns: use of silver nitrate and of isolators, *Brit. Med. J.,* **ii,** 1288.

Colebrook, L. (1950). *A New Approach to the Treatment of Burns and Scalds,* p. 122, Fine Technical Publications, 39 Wilton Rd, London, S.W.1.

Haynes, B. W. and Hench, M. E. (1966). Total hospital isolation: concept and practice. *Second International Congress on Research in Burns,* edited by A. B. Wallace and A. W. Wilkinson, P. 550, Edinburgh.

James, K. W., Jameson, B., Kay, H. E. M., Lynch, J. and Ngan, H. (1967). Some practical aspects of intensive cytotoxic therapy, *Lancet,* **i,** 1045.

Levenson, S. M., Trexler, P. C., La Conte, M. and Pulaski, E. J. (1964). Application of the technology of the germ-free laboratory to special problems of patient care, *Amer. J. Surg.,* **107,** 710.

Lidwell, O. M., Polakoff, S., Jevons, M. P., Parker, M. T., Shooter, R. A., French,

V. I. and Dunkerley, D. R. (1966). Staphylococcal infection in thoracic surgery: experience in a subdivided ward, *J. Hyg., Camb.*, **64,** 321.

Lowbury, E. J. L. (1954). Air conditioning with filtered air for dressing burns, *Lancet*, **i,** 292.

Lowbury, E. J. L. (1960). Infection of burns, *Brit. med. J.*, 994.

Lowbury, E. J. L. (1965). Research on control of hospital infection by air-conditioned isolation wards, *Nursing Mirror*, 17th September.

Report (1965). Isolation requirements for surgical units, by a Working Party from the North East Region Pathology Advisory Committee, *Lancet*, **ii,** 895.

Robertson, A. C., Lynch, J., Kay., H. E. M., Jameson, B., Guyer, R. J. and Evans, I. L. (1968). Design and use of plastic tents for isolation of patients prone to infection, *Lancet*, **ii,** 1376.

Woodruff, M. F. A., Nolan, B., Bowie, J. H. and Gould, J. C. (1968). The Nuffield Transplantation Surgery Unit, *Lancet*, **i,** 905.

TABLE 1

Bacterial counts on $3\frac{1}{2}$-inch settle-plates after six hours' exposure

Site of sampling	*Mean bacterial counts*	*Number of observations*[a]
Open ward air	46·1	11
Isolator with filters	0·08	5
Isolator with no filters	1·19	8

[a] Each observation represents a sampling with a number of settle-plates on one day.

TABLE 2

Bacterial counts of air by Andersen sampler

	Total bacteria per cubic foot of air		
Site of sampling	*Experiment 1 (quiet ward)*	*Experiment 2 (busy ward)*	*Experiment 3 (quiet ward)*
Open-ward air	2·4	7·3	2·3
Isolator with filters	0·2	0·01	0·03
Isolator with coarse filter only	—	0·5	0·22
Isolator with no filter	1·2	—	0·13

TABLE 3

Distribution of particle sizes of airborne bacteria in open ward and in isolator

Estimated particle size	Numbers of bacteria-carrying particles in air[c]							
	Open ward		Isolator with coarse filter only		Isolator with no filter		Isolator with filters	
	Expt. 1[a]	Expt. 2[b]	Expt. 1	Expt. 2	Expt. 1	Expt. 2	Expt. 1	Expt. 2
9·2 μ and above	15 (1 *Staph. aureus*)	145	—	0	1	—	1	0
5·5–9·2 μ	15 (1 *Staph. aureus*)	109	—	1	3 (1 *Staph. aureus*)	—	2	0
3·3–5·5 μ	19	79	—	0	5	—	0	0
2·0–3·3 μ	11	63	—	7	6	—	2	0
1–2 μ	11	42	—	22	19 (2 *Staph. aureus*)	—	0	0
Less than 1 μ	0	0	—	0	1	—	1	0

[a] Ward quiet. [b] Ward busy.
[c] Per 30 cubic feet in Experiment 1, per 60 cubic feet in Experiment 2. Counts of *Staph. aureus* were made in Experiment 1 only.

TABLE 4

Controlled trial of isolators: infection of burns with Pseudomonas aeruginosa
(interim assessment, July 1969)

Bacteriology	*Phase of study*	*Patients*				
		in isolators with coarse and fine filters	*in isolators with coarse filters only*	*in isolators with no filters*	*in isolators (total)*[a]	*not in isolators*[a]
Ps. aeruginosa	1	2	2	–	4	7
Acquired	2	0	–	0	0	4
Ps. aeruginosa	1	9	7	–	16	2
Not acquired	2	6	–	5	11	2
Total	–	17	9	5	31	15

[a] $X^2 = 13{\cdot}5$, $P < 0{\cdot}001$.

TABLE 5

Controlled trial of isolators for burns: infection of burns with Staph. aureus *resistant to two or more antibiotics (RR) (interim assessment, July 1969)*

Bacteriology of burns in phases 1 and 2	Patients				
	in isolators with fine and coarse filters	*in isolators with coarse filters only*	*in isolators with no filters*	*in isolators (total)*	*not in isolators*
Staph. aureus RR first isolated in					
1st week	8	3	2	13	11
2nd week	3	0	0	3	2
Later	1	2	0	3	0
Total	12	5	2	19	13
Staph. aureus RR not found	5	4	3	12	2
Total	17	9	5	31	15

TABLE 6

Controlled trial of isolators for burns: nasal acquisition of Staph. aureus *resistant to two or more antibodies (RR) (interim assessment, July 1969)*

Bacteriology of nose in phases 1 and 2	*Patients*				
	in isolators with coarse and fine filters	*in isolators with coarse filters only*	*in isolators with no filters*	*in isolators (total)*	*not in isolators*
Staph. aureus RR first isolated in					
1st week	6	2	2	10	9
2nd week	4	1	0	5	2
Later	0	0	0	0	0
Total	10	3	2	15	11
Staph. aureus RR not found	7	6	3	16	4
Total	17	9	5	31	15

TABLE 7

Controlled trial of isolators for burns: comparability of groups (interim assessment, July 1969)

	Group 1. Isolators with filters	*Group 2. Isolators with coarse filter only or with no filter*	*Group 3. No isolator*
Number of patients	17	14	15
Mean age (years) and range	$8\frac{1}{2}$ (2–27)	10 (2–28)	$4\frac{1}{2}$ (1–12)
Mean area of burn and range	14% (4–30)	11% (7–19)	15% (5–30)
Number treated by covered method	12	9	13
Number treated by exposure	4	2	1
Number treated by covered and exposure methods	1	3	1

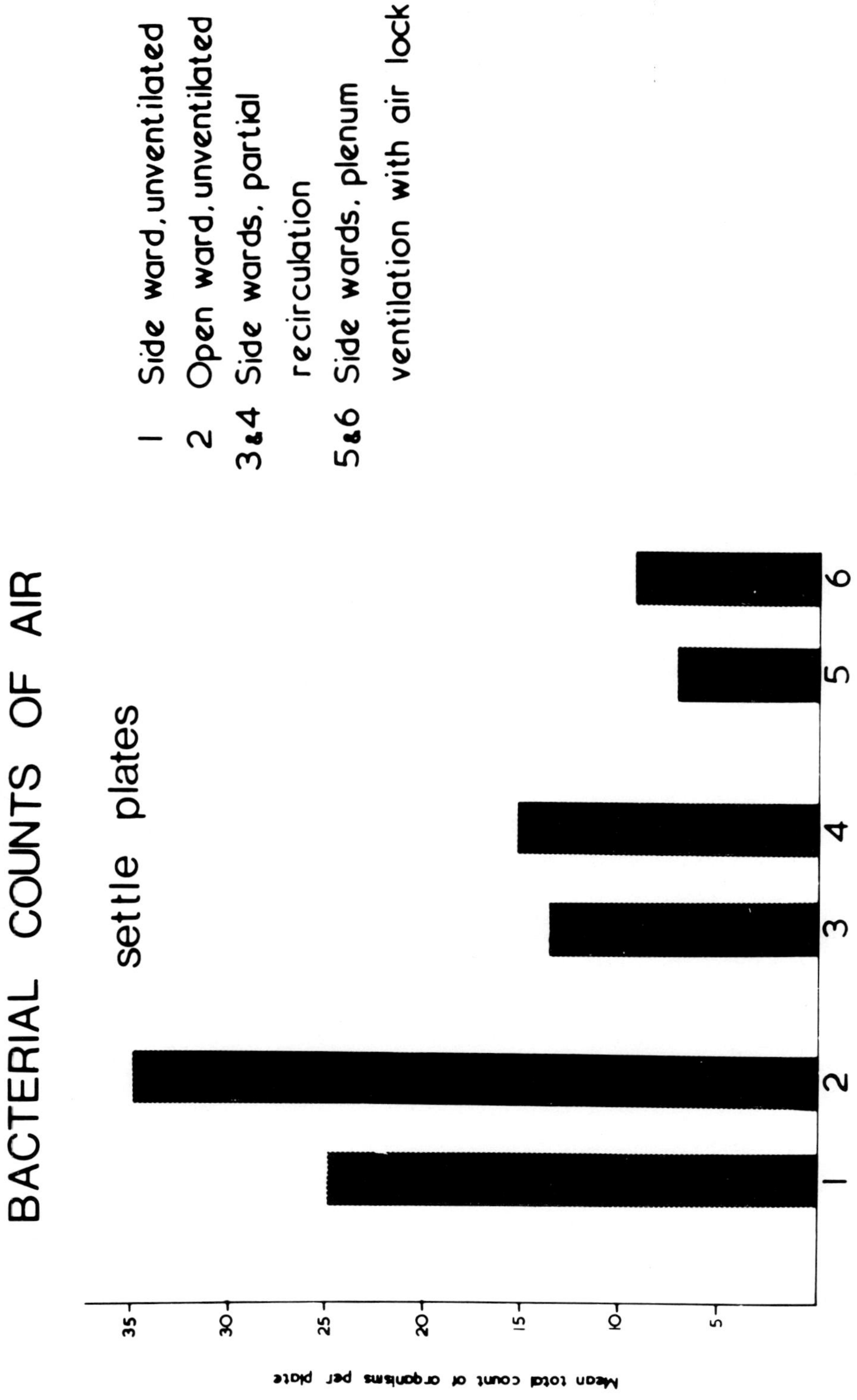
BACTERIAL COUNTS OF AIR
settle plates
1 Side ward, unventilated
2 Open ward, unventilated
3 & 4 Side wards, partial recirculation
5 & 6 Side wards, plenum ventilation with air lock
35
30
25
20
15
10
5
Mean total count of organisms per plate
1
2
3
4
5
6

Figure 1

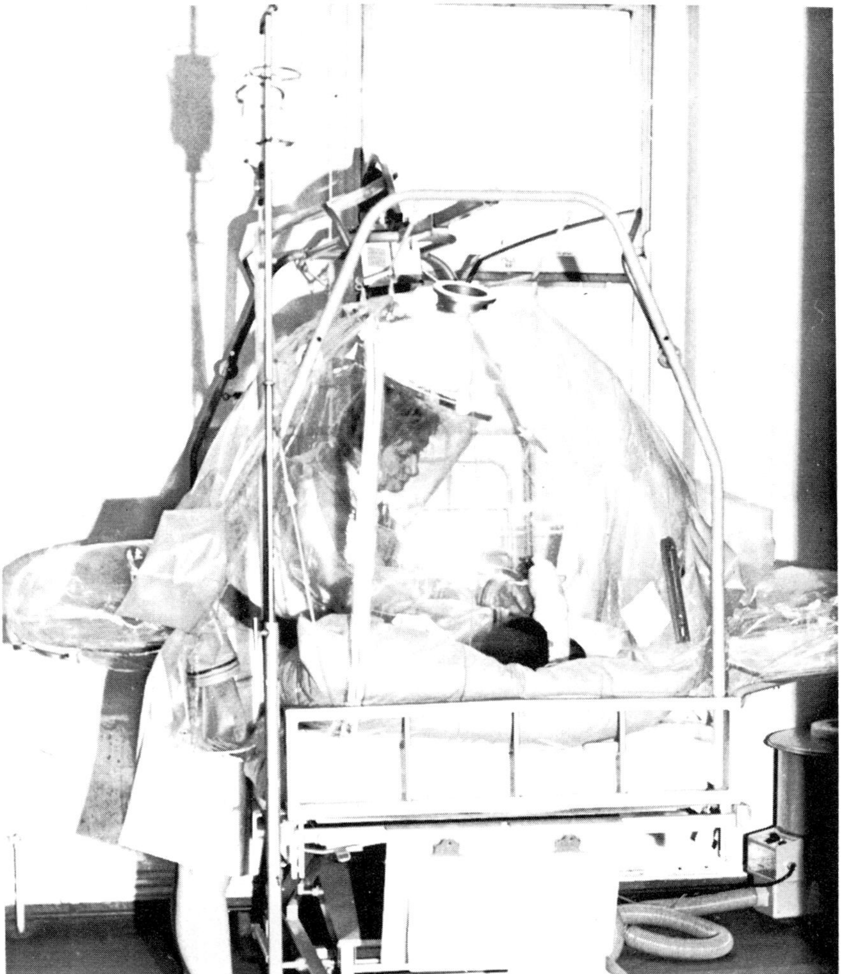

Figure 2

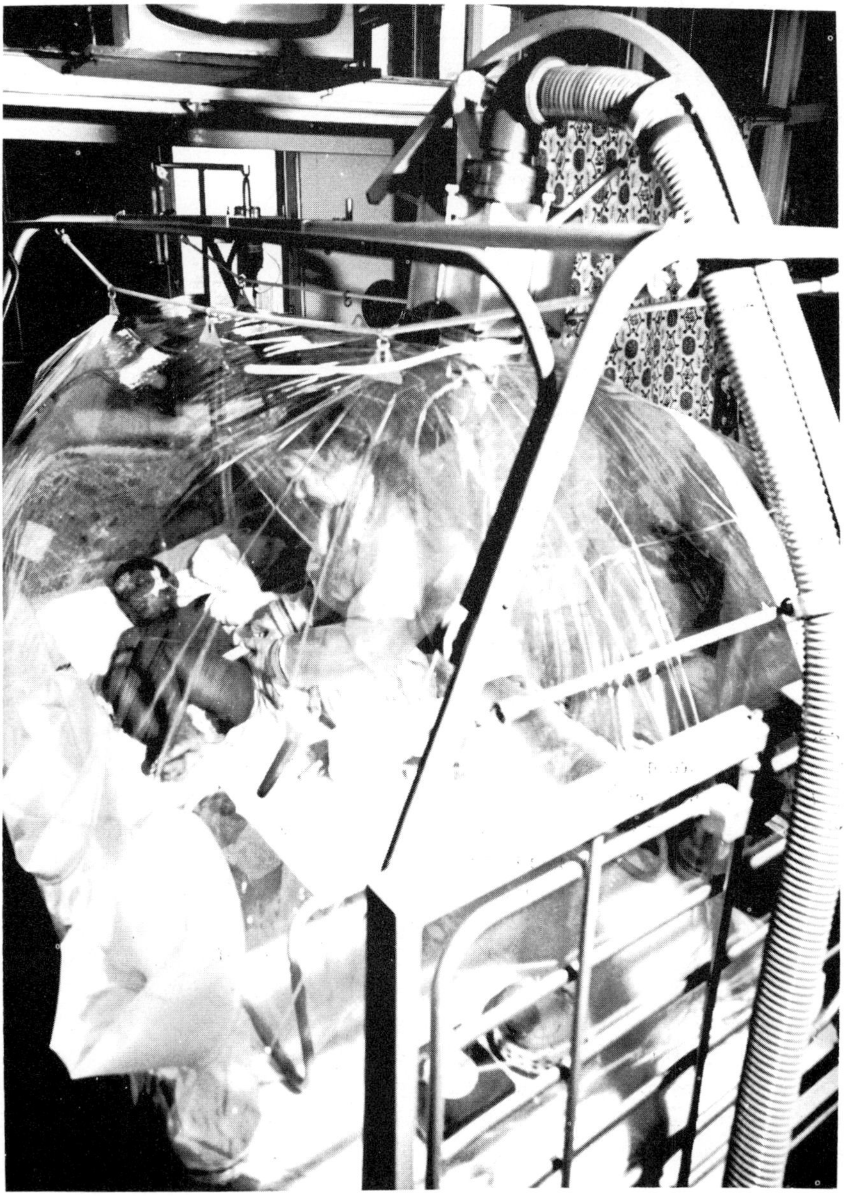

Figure 3

The control of airborne contamination in sterile areas

G. Sykes

Microbiology Division, Boots Pure Drug Co. Ltd., Nottingham

Sources of contamination

Microbial contamination is always transmitted either by direct contact with a source already contaminated, or by an airborne route, and it is derived from the operators, from the equipment and containers in which the material is being handled, or from the air. In sterile filling rooms conscious attempts are made to eliminate the sources of infection by sterilizing the material itself, by sterilizing the equipment and containers, by removing organisms from the ventilating air and by isolating the operators as far as possible from direct contact with the work in hand. Even so there is always some hazard of infection by airborne routes arising simply from the presence of human operators, and to some extent the two are interlinked.

There are well-established methods for sterilizing the air entering an enclosed space, be it a screen, a glove box or a large room, but the problem of dealing with contaminations arising from the operators is much more difficult, largely because it is an *in situ* phenomenon. In fact, in a properly designed and controlled sterile area the operator is the major source of infection and, even worse, a continuous one. The simplest of movements gives rise to frictional action between the clothing and the skin, and this in turn causes skin scales and their attendant micro-organisms to be shed from the body surfaces. Many of these particles are relatively large and so settle out fairly rapidly—and in so doing contaminate any surface on which they may alight—but the smaller particles can remain airborne for quite long periods. Therefore we have a situation in which it is theoretically possible to prepare and maintain a room and its contents virtually sterile, but which as soon as it is occupied becomes progressively more heavily contaminated with human-borne organisms, first in the air and then on the various horizontal surfaces. The extent of the contamination obviously depends on the number of persons in the room and the amount and duration of their various activities.

Prevention of contamination. Isolating the individual

In an aseptic working area the problem can be met in one of two ways: by isolating either the individual from the job or the job from the individual. The first solution involves clothing the operator in such a manner that the organisms he sheds are prevented as far as possible from being released into the air and so from gaining access to the work in hand. Adequate ventilation of the area with sterile air is also necessary to carry away the organisms which do escape, to prevent them from subsequent settling.

Much useful information has resulted from several investigations (e.g. Bethune, Blowers, Parker and Pack 1965; Bernard, Spears, O'Grady and Shooter 1965; Blowers and McClusky 1965) on the effects of clothing on the shedding of organisms. In general more organisms are dispersed from below the waist than from above it, and skirts allow more organisms to escape than do trousers (by a factor of about five), although men disseminate more organisms than do women. In this context, the extent and vigour of the activity taking place is important, as illustrated in some findings quoted in Table 1. In these experiments, made in a room of about 850 $ft.^3$ capacity with fans in the room to keep the air in constant movement, only one operator was in the room at a time, but the results were obtained with two separate operators. On each occasion the operator wore either his ordinary laboratory coat or he was fully covered with a sterile head cover, sterile cotton gown, trousers, canvas overshoes and rubber gloves. The routine followed was: the operator first entered the room and sat still for 3 minutes, then he stood swinging his arms for 3 minutes, next he walked around the room for 3 minutes and finally exercised more violently for 3 minutes. During these 4×3 minutes periods samples of 1 $ft.^3$ of air were taken at regular intervals by means of a slit sampler. The figures quoted show that the traditional sterile clothing is of but little value in restricting the release of organisms into the air, and some further experiments confirm this. In these the dress varied, but the activity of the operator was kept constant and the results obtained are quoted in Table 2. Besides the relative inefficiency of the standard aseptic dress, the apparent difference between cotton material and nylon is interesting.

Real improvements are achieved only when impermeable dustproof dress is worn, as demonstrated by Bernard *et al.* 1965, who obtained by this means reductions of 90 per cent or more of the organisms disseminated. Further reductions occurred when the gown was replaced by a one-piece suit, and again when the feet were totally enclosed in plastic bags and the trouser legs sealed with cycle clips or rubber bands (Table 3). The figures quoted were obtained in small chambers ventilated by means of filtered air from fans in the ceiling and with one operator carrying out various 'exercises'.

Isolating the job

Our experience indicates that the method of isolating the individual from the job, although of some value in reducing the incidence of contamination of sterile material, has a number of disadvantages, not the least of which is the discomfort of the operator. Much better and more convenient is the alternative method of isolating the job from the individual, the method we have practised in our organization for many years.

In this method the work in hand is enclosed in a suitably designed screen which is fed with a continuous supply of sterile air. The air is introduced through a filter sited preferably at one end of the screen so that it flushes the whole of the screen evenly in a laminar style and then passes out through an aperture at the other end of the screen. From experience an air-change rate of at least 1/min. is necessary to keep the airborne count within the screen at a satisfactory low level. It is also preferable for the air flow to pass first over the immediate working site and then towards the operator, rather than in the opposite direction.

The only access the operator has to the screen is via arm holes cut suitably in the front wall through which he puts his gloved hands to carry out the necessary manipulations. The arm holes are fitted with detachable and sterilizable sleeves which grip the gloves at the wrist, thus isolating completely the work in hand from the operator. With such a system it is not necessary for the whole room to be treated with sterile air, neither is it necessary for the operators to wear sterile coverings. But clearly such precautions provide additional safeguards, and as such are desirable.

Methods of assessing airborne contamination

Airborne contamination is assessed either by the so-called 'settling-plate' method, using exposure times ranging from a few minutes to one or two hours, or by means of the slit sampler or similar apparatus, in which all of the particles down to a given size in a known volume of air are collected and cultured. The settling-plate method is entirely empirical and the counts so obtained, besides depending on the nature of the activity taking place in the room, will vary also with the exact siting of the plate in relation to the work going on. To demonstrate this, counts were taken simultaneously at bench and floor level in a suite of rooms each ventilated with 15 changes of air/hour: some slit-sampler counts at bench level were also taken at the same time. The bench-level counts were taken near a sterile filling screen, and the floor-level ones usually near to where the operators were sitting: typical results are given in Table 4.

The settling plate does not assess the numbers of falling organisms in the strictly gravitational sense, but rather those which are impinged on and adhere to the agar surface, and this in turn depends on the velocity of the air flow over the plate, the angle of the air flow relative to the agar surface, the mass of each airborne particle and the retentiveness or adhesiveness of the surface. In a similar manner, the slit sampler also gives somewhat empirical values because, although it assesses the total number of organisms in the air, many of the particles are so small that in still air they would be unlikely to settle out within a reasonable period of time, and in a ventilated space the majority of them would be swept away and thus present no hazard. To illustrate: according to Stokes' law the mean settling time in still air from a $1/\text{ft.}^3$ column of air above a $2\frac{1}{2}$-in. Petri dish is about 17 hours for particles of

1 μ diameter and about 7 mins. for particles of 5 μ diameter. Such particles, therefore, would not often settle onto a sterile surface, liquid or solid, simply by gravitational fall. The greatest hazard is from moving air carrying organisms in the direction of, or onto, the sterile surface. This further underlines the importance of arranging the work in the sterile area so that potential sources of contamination are, as far as possible, downwind to the immediate site of aseptic operation: this applies particularly to the air-flushed screen.

Comparisons of the settling-plate and slit-sampler methods

In practice, when the two samples are taken in close proximity, the count/ft.^{3} of air by the slit sampler is about the same as that from a settling plate exposed for 15 minutes. This is a purely arbitrary relationship and obviously will vary under different conditions. It would not obtain, for example, in a room in which the air-change rate is high or in which there is considerable turbulence.

A feature of all airborne count determinations is the range of values obtained under apparently the same conditions, and the considerable widening of the ranges as the exposure period is lengthened (Figure 1). This can only be explained on the assumption that in some instances a flurry of activity occurred at some stage during the exposure period which increased temporarily the airborne count, thus giving a higher than expected total count for the whole period. In the settling-plate method such an occurrence would not be seen directly, but it accounts for the unevenness of distribution of the counts about the median and average counts. A slit-sampler plate with its minute-to-minute records would show this more clearly, and exactly when the flurry occurred.

Viable airborne counts in ventilated sterile rooms

In our organization all sterile filling of injections, etc., is carried out in rooms flushed with sterile air and maintained at a slightly positive pressure. The air is introduced through a continuous line of filters situated at ceiling height on one wall of the room and extracted near ground level at the opposite wall; the air-change rate is 15/hour. The direction of the air flow is therefore roughly downwards diagonally across the room. In one such room, measuring 30 $\times$ 60 feet, viable air counts were made at intervals during one day, at one point about 4 feet from the air input filters and at another about 4 feet from the extract filters. The number of people in the room during the experiment varied up to 20. The viable counts obtained/5 ft.^{3} of air are shown in Figure 2.

These findings are similar, although, as expected, at a higher level, to those of Favero, Puleo, Marshall and Oxborrow (1966) in industrial clean rooms, one example of which is illustrated in Figure 3. These counts were taken in a room of capacity 1080 ft.^{3} supplied with laminar flow air at a change rate of 165/hour and with up to eight operators present. The air entry counts (site 1) were taken at a point near the centre of the input filter wall, and the extracted air counts at a point near the opposite wall.

Organisms on inert surfaces

So far consideration has been given only to the fate of organisms arriving on nutrient surfaces. But what of those on inert surfaces such as glass and metal? One might expect there to be a gradual and continuous accumulation of organisms. albeit of the genera more resistant to drying, but experience shows it not to be the case. Favero *et al.* (1966) quote several references to the 'plateau' phenomenon which occurs with organisms settling on glass and stainless steel strips. In this it is always found that the microbial count/unit area increases during the first week, after which it remains constant for an indefinite period. In a preliminary investigation of this in our own laboratories we exposed glass microscope slides, stainless steel and plastic strips in a clean bacteriology laboratory in which there is an air-change rate of eight per hour; the relative humidity during the experiment ranged from *c.* 55 to 70 per cent. At intervals up to 28 days the bacteria count was assessed on each of two of the different types of strips by transferring an individual strip to a Petri dish, pouring over it molten agar, swirling the agar for a few seconds and then incubating at 32°C. The average count (Table 5) showed considerable day-to-day variations, and there was a slight tendency for the plastic to yield higher counts than did the metal or glass surfaces. Much more important was the fact that the counts after two days exposure were roughly the same as those for longer periods up to 28 days, thus indicating that a 'plateau' value had been reached even within two days.

Accepting the simple explanation offered by various authors that on any surface there is always a natural die away of organisms which at some level equals that of the rate of deposit of organisms, thus establishing the plateau phenomenon, one might expect that the residual viable population would consist mainly, if not exclusively, of the most resistance types of organisms, i.e. the bacterial spores. In the experiment quoted above we found consistantly that 60–70 per cent of the colonies were of cocci and 30–40 per cent of G_+ spore-forming rods; in addition there were mould colonies constituting usually 10–15 per cent of the total bacterial colonies. These findings are in agreement with those of Favero *et al.* (1966), who recorded the survival of substantial numbers of micrococci and coryneform bacteria as well as mould and bacterial spores.

To test further the stability of these settled organisms, glass slides after three days' exposure in the laboratory were transferred individually to Petri dishes which were then closed and kept in the laboratory. At intervals the viable bacteria and moulds on each of four slides were determined by the method already described, but using ordinary nutrient agar, with incubation at 32°C for determining the bacterial count and glucose agar, with incubation at 25°C for the mould count. These results (Table 6) showed a rapid decline in the viable bacterial population during the first 24 hours, followed by a more steady decline during the remaining period. The mould population was more persistent and reached an apparently steady level after about seven days. The surviving bacteria after the first 24 hours were all G_+ spore-forming rods. These findings differ from those of Lidwell and Lowbury (1950), who found that organisms on dust swept from ward floors, which may well be a special situation, were much more persistent. The subject clearly merits further investigation.

REFERENCES

Bernard, H. R., Speers, R., O'Grady, F. and Shooter, R. A. (1965). *Lancet*, **i,** 458.

Bethune, D. W., Blowers, R., Parker, M. and Pack, E. N. (1965). *Lancet*, **i,** 480.

Blowers, R. and McClusky, M. (1965). *Lancet,* **ii,** 681.

Favero, M. S., Puleo, J. R., Marshall, J. H. and Oxborrow, G. S. (1966). *Appl. Microbiol.,* 14, 539.

Lidwell, O. M. and Lowbury, E. J. (1950). *J. Hyg., Camb.,* **48,** 6.

TABLE 1

Dissemination of organisms during various activities

	Average viable count/cu. ft. of air with	
Activity	*Normal dress*	*Sterile dress*
Control	< 1	< 1
Sitting quietly	3 (0–7)	2 (0–4)
Swinging arms	32 (7–73)	13 (1–41)
Walking	55 (25–75)	35 (4–75)
Running on the spot	142 (94–201)	65 (19–128)

Figures in brackets are the ranges of counts.

TABLE 2

Effect of dress on the dissemination of skinborne organisms

	Airborne count/ft.^{3}	*Total no. of organisms dispersed in* $8\frac{1}{2}$ *min.*
1. Shirt sleeves	160 (40–244)	135,000
2. Laboratory coat (clean or dirty)	55 (36–75)	46,000
3. Sterile cotton tunic (side buttoned), with trousers, overshoes etc.	16 (2–28)	13,000
4. As 3, but in nylon	35 (4–65)	30,000
5. As 3, but long gown (tape fastened at back)	34 (4–75)	28,000

The figures in brackets are the ranges of 6 × 1 ft.^{3} samples taken over a period of eight minutes.

TABLE 3

The escape of skinborne organisms through various types of dress (after Bernard et al. *1965)*

Dress	*Airborne count/ft.*^{3}
1. Surgical gown	4–400
2. Surgical gown, with feet enclosed in polythene and trouser legs sealed at the ankles	2–200
3. As in 2 but gown replaced by closely woven cotton one-piece suit	1–30
4. As in 2 but gown replaced by one-piece suit made with impermeable material	about 1

TABLE 4

Airborne counts at different points in sterile rooms

Room no.	*Average settling-plate count (15 min.) at*		*Average slit-sampler count/ft.*^{3} *at bench level*
	Bench level	*Floor level*	
1	1·5	4·1	2·0
2	1·8	5·5	2·0
3	1·1	3·7	1·2

TABLE 5

Numbers of bacteria on different surfaces after exposure for various periods

Period (days) of exposure	No. of bacteria[a] found on Glass slides	Steel strips	Plastic strips
2	25	16	57
5	22	15	18
7	13	8	29
14	13	6	18
21	21	7	24
28	39	22	28

[a] The counts quoted are averages from two strips.

TABLE 6

The survival of micro-organisms on glass surfaces

Period (days) of storage	Total[a] count of	
	Bacteria	Moulds
0	77	18
1	12	17
2	3	12
3	5	9
4	0	8
7	5	5
11	1	7
18	1	5

[a] The totals are from four glass slides on each occasion.

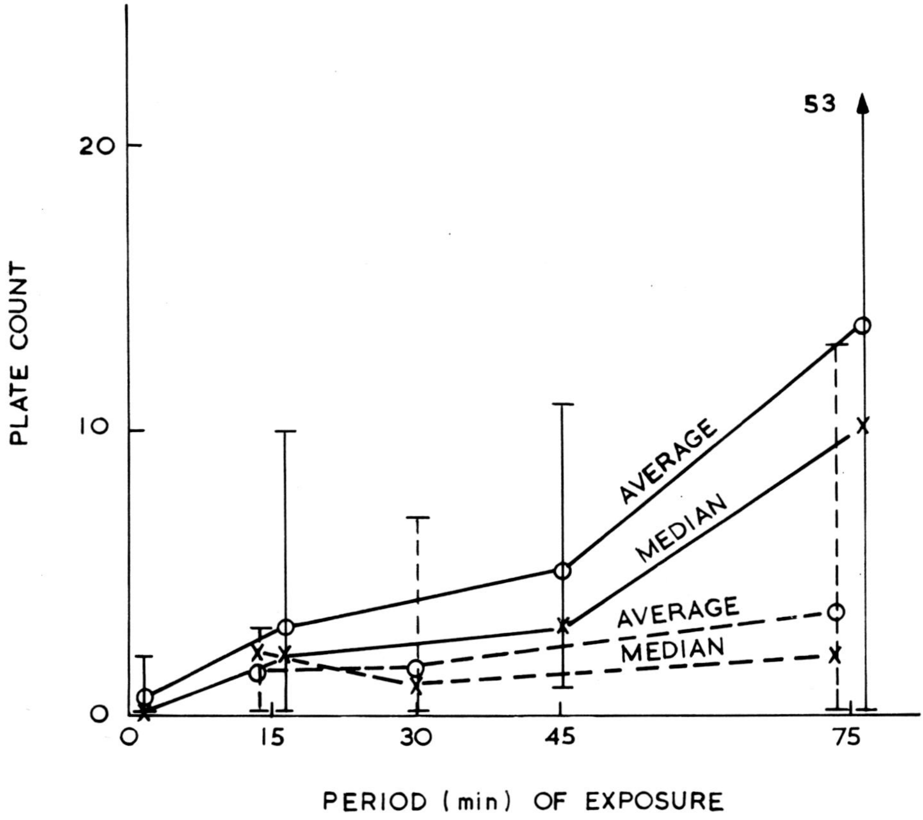

Figure 1. Variations of settling-plate counts with period of exposure. Vertical lines, range of counts; broken lines, at bench level; solid lines, at floor level. Circles, average counts; crosses, median counts.

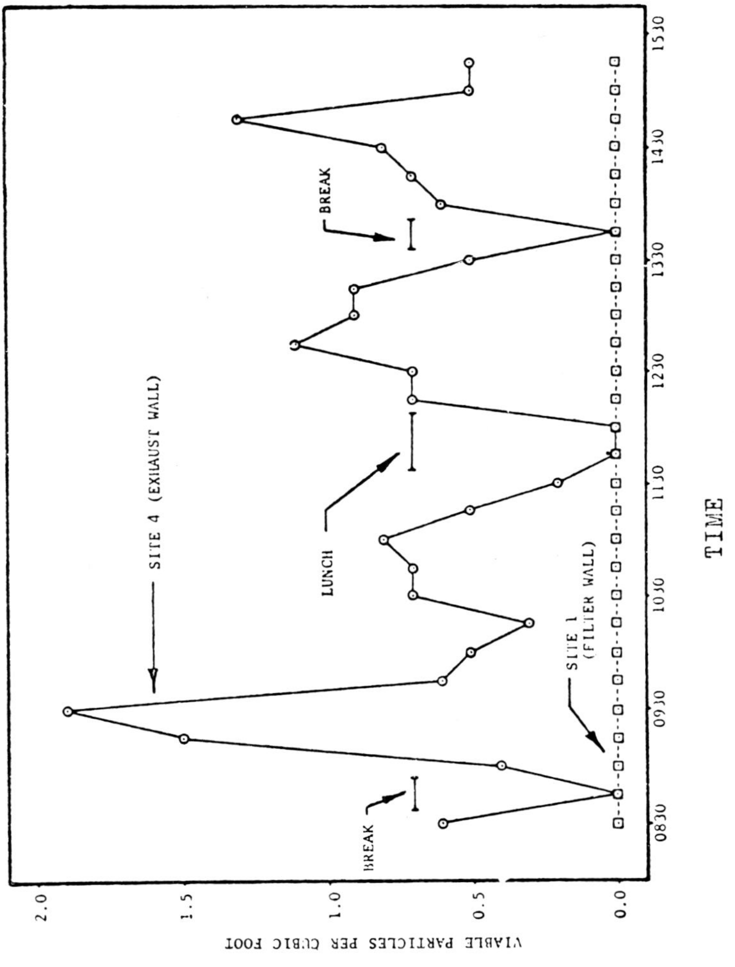
SITE 4 (EXHAUST WALL)
SITE 1 (FILTER WALL)
LUNCH
BREAK
BREAK
VIABLE PARTICLES PER CUBIC FOOT
2.0
1.5
1.0
0.5
0.0
0830
0930
1030
1130
1230
1330
1430
1530
TIME

Figure 2

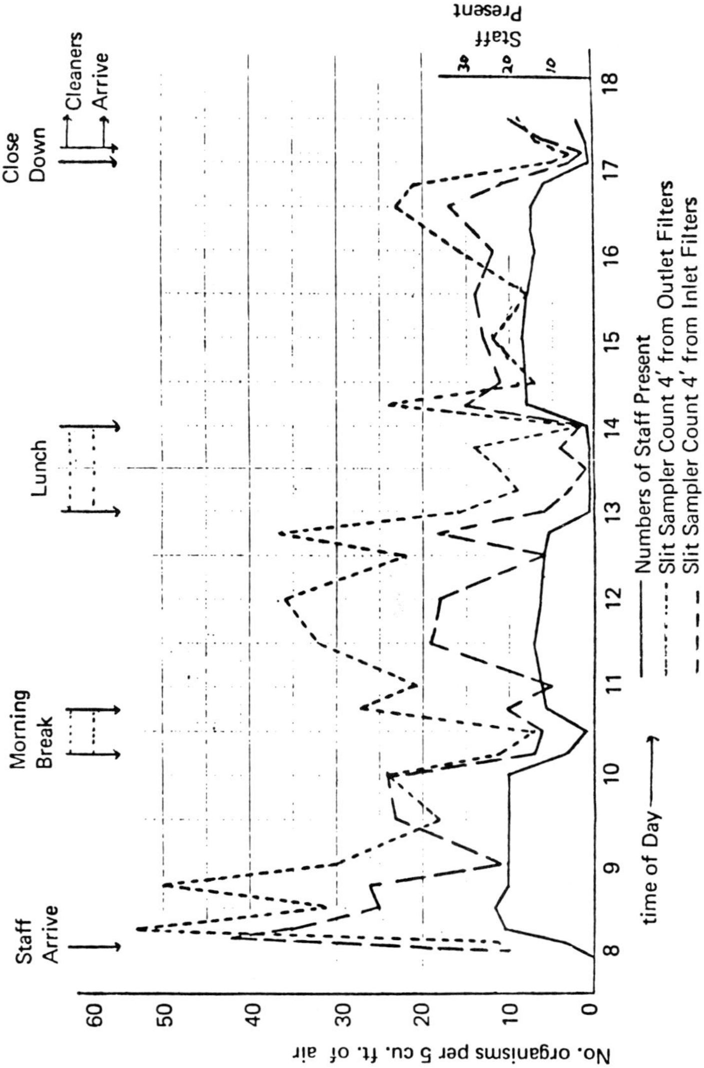
Staff Arrive
Morning Break
Lunch
Close Down
Cleaners Arrive
No. organisms per 5 cu. ft. of air
60
50
40
30
20
10
0
8
9
10
11
12
13
14
15
16
17
18
Staff Present
30
20
10
time of Day
Numbers of Staff Present
Slit Sampler Count 4' from Outlet Filters
Slit Sampler Count 4' from Inlet Filters

Figure 3

The clinical value of a bacteria-free patient-care unit

John F. Burke, M.D.

Massachusetts General Hospital, Boston, U.S.A.

The classic methods of preventing the spread of infectious disease in a hospital suffer from two disastrous flaws. First, they are economically impractical because they require the inefficient use of medical personnel, patient care, equipment, and hospital beds and floor space. Second, and perhaps most important, they are not effective. Hospital practice of today has evolved to the point where infection usually acquired within the hospital is more often the cause of the patient's death than the disease for which the patient was admitted. This fact has long been recognized in the case of thermal burns. It is true also for the developing area of allotransplantation and, although not as clearly seen, it is also becoming true on the general medical and surgical wards, for if a careful examination of hospital patients is carried out, it will be seen that a sizeable portion of deaths result from in-hospital acquired infections which will frequently begin in the lung or the urinary tract of a debilitated patient. This in-hospital cross-infection is becoming a more alarming source of clinical problems, not only because of the ineffective methods used to prevent bacteria from an infected patient from spreading disease to his fellows, but also because an ever-increasing proportion of the hospital patient population have a serious defect in host-defence against bacterial infection produced either by disease or by therapeutic design. The efficient solution to the problems of cross-infection in a hospital, therefore, cannot be solved by a method which simply prevents bacterial spread from the infected patient to his neighbor, but must also contain a design which will defend a debilitated patient from the usual levels of environmental bacteria when his host's defences have been temporarily destroyed. Therefore it is evident, at least from the economic point of view, that the most successful solution will be one which solves both forward and reserve isolation problems using a single method.

The experience gained through the use of current procedures of patient isolation has demonstrated that at best the use of a single room for patient housing and protective clothing for medical personnel immediately reduces the number of bacteria transferred but does not prevent cross-contamination. With this experience in view, a number of systems have recently evolved which provide an impermeable barrier between the patient and the rest of the hospital environment. These methods have successfully prevented the transfer of bacteria from the infected patient to the environment as well as protecting the deficient patient from ambient environmental bacteria. Although they are highly effective in preventing cross-infection, they have, on the whole, not been clinically successful for a number of reasons. First, they do not prevent the endogenous spread of infection from one area of the patient to another; second, nursing and medical care is made cumbersome and at times impossible; third, they place a considerable psychological burden on an already harried patient; fourth, they require specially trained, expert personnel and a considerable increase in equipment and storage space; and fifth, they are too expensive to be generally used.

The system to be described in this paper has been an attempt to solve the problems of forward and reverse isolation and to reduce the risk of auto-infection through the use of a single unit. Attempts have been made to simplify the delivery of medical care without nurse or doctor entering the bacteria-free patient environment, to reduce the psychological hazards of isolation, and to provide a piece of portable equipment of moderate cost. The unit itself has been developed over the past four years by a multidisciplinary team of engineers, bacteriologists, and physicians. Six of these bacteria-free patient-care units have been in continuous use beginning in December of 1968 in the Shriners Burns Institute in Boston.

Air-handling and bacteria-removal design

The research efforts which have been carried out have led to the development of two types of air-handling and bacteria-removal systems and a single patient-care space. These systems are now being clinically evaluated and attempts to improve nursing efficiency are being explored. From the engineering standpoint, two mechanical systems are described by the functional block diagrams of Figure 1. Configuration A of Figure 1 uses HEPA filters (0·3 micron mean pore size) and provides an average chamber air velocity of 50 feet per minute with 85 per cent recirculation. This is the type of mechanical system permanently installed in the Shriners Burns Institute. The second type provides a portable unit shown in configuration B in Figure 1, employing electrostatic filters and 2537 Å ultraviolet radiation to eliminate viable particles. This system also recirculates 85 per cent of air with a chamber velocity of 35 feet per minute.

The performance characteristics of the configuration A unit are being evaluated by a continuing program of viable particle sampling within the chamber (occupied and unoccupied) as compared with the ambient level in the surrounding ward area. The HEPA filter performance is specified to eliminate 99·97 per cent of the particles in excess of 0·3 micron diameter with each pass-through the unit. Volume flow of air in the unit is 300 cubic feet per minute, of which about 450 cubic feet per minute are not recirculated. With ambient air containing less than 10 viable particles per cubic foot, one would expect the unoccupied chamber to contain less than five viable particles per 10,000 cubic feet of air in the steady state condition. This has, in fact, been confirmed by the air-sampling program. These units utilize vertical down-flow with ceiling plenums comprising the chamber ceiling area. Since full ceiling plenums are not employed and no special provisions are made for preventing standing eddy patterns in the chamber, air flow is not uniform throughout the unit. In spite of this, the particle count has never exceeded 0·06 viable particle per cubic foot in the occupied unit even during the period of bed linen and dressing change. Usual viable particle count is on the order of 0·04 per cubic foot of chamber air above the patient level.

The mechanical system using configuration B of Figure 1 contains electrostatic filtration which is 99·9 per cent effective in removing 5 μ or larger particles. Particles smaller than 5 μ may be efficiently sterilized with ultraviolet radiation. In this system, $4{\cdot}5 \times 10^5$ microwatt per second per centimeter square of 2537 Å radiation is provided, inhibiting 99·9 per cent of colony formation. Viable particle removal achieved in this manner is comparative with the HEPA unit, but has the considerable advantage in that the pressure drop across the bacteria-removal system is in the order of one magnitude lower than in the HEPA filter configurations. This unit has full ceiling and floor plenums and has been designed to provide a uniform piston air flow throughout the chamber with no outstanding eddies and a minimum of local turbulence.

Description of bacteria-free patient-care chamber

The patient-care space itself is a 6×10 foot rectangular area surrounded by a transparent polyurethane enclosure. The space contains the patient's bed, and if required a bedside table or chair. The entire 6×10 foot area is subjected to ceiling to floor air flow at the rate of 50 linear feet per minute. Air entering into the unit is filtered at the ceiling plenum via sealed HEPA filters and enters the patient-care environment essentially bacteria-free. The air washes over the patient and his surroundings and is exhausted through a floor plenum to be partially recirculated by way of a course filter and blower system as described above. A variable amount of sterile air is exhausted immediately below the ultra filters, and an equal amount is added at the blower to prevent CO_2, heat and water vapour build-up in the unit. Using 85 per cent recirculation, this can be effectively accomplished as demonstrated in Figures 2 and 3. Each chamber is provided with an individual blower system remotely located from the chamber itself. The remote location of the blower and the use of acoustical baffling provide a patient-care environment that is without air or mechanical noise. Sound evaluation tests have demonstrated that there is no increase in the ambient ward noise level when all six units are turned on at once. It has been impossible for the patients housed in the chamber to tell whether the air flow is on or off.

The curtain walls surrounding the 6×10 foot space are constructed of thin transparent sheets of polyurethane, allowing the patient an unobstructed view of the surroundings and the nursing staff an unobstructed view of the patient. In order to facilitate medical care, four of the units are placed together in a right-angle configuration with two units on each limb of the angle. The grouping of units together allows a nurse and aide to care for these four patients as on an open ward, with almost the same efficiency as if isolation precautions were not being carried out. The one-to-one patient to nurse ratio which is so often required for the proper care of a critically ill patient in classic isolation is therefore not required. Each unit now in use at the Burns Institute is surrounded with a moveable opaque cloth curtain of the variety found in multiple-occupancy hospital rooms, assuring patient privacy when this is required.

Further to facilitate the delivery of nursing care, a system has been devised whereby medical treatment can be delivered without entering the patient-care chamber itself. In this way, all protective clothing is eliminated except for the hands and arms themselves. This is accomplished through the use of specialized end and side curtain walls. The end wall contains a pass-through air lock for material and equipment taken into or out of the unit. The end wall also contains a folding air lock to allow properly dressed personnel to enter the unit if this is necessary. The two side walls are constructed of thin transparent polyurethane sheets held under tension by polyurethane cords to improve the transparent character of the sheets. In order to provide indirect patient access, each side wall is made of two panels (an upper and a lower). These panels overlap each other 4 feet from the floor with overlap of 8 inches. The upper forms the outer layer of the overlap, and the edges of both the upper and the lower panel are held taut but moveable by elastic polyurethane cords. Access is gained to the interior of the chamber by donning shoulder-length, sterile polyurethane gauntlets. The overlap portion of the wall panels is then spread, and without touching either panel, the hands and arms are put into the chamber. The elastic edges of the panels are then allowed to close about the upper arms. Since the overlap runs from one end of the unit to the other, the nurse or doctor delivering patient care can move from one end of the unit to the other without interruption. Both side curtain walls are constructed of the same type of overlapping panels for access, but because of the width of the unit, both curtain walls cannot be conveniently close to the patient to provide efficient nursing care. In order to allow ease of operation and to allow nursing care to be carried out on both sides of the patient's bed at the same time, if necessary, the bed is placed adjacent to one of the curtain walls, while the opposite curtain wall is constructed so that it can be moved into the unit by a distance of 2 feet. The moveable access wall is housed in the extended position except when pushed to the patient's bedside when required. Using this arrangement, most of the nursing and medical care can be carried out through the access wall adjacent to the patient's bed. If ambulatory, therefore, the patient can move about freely in the space on the opposite side or front of his bed. If, however, complex treatment must be carried out, such as burn dressings or minor surgical procedures, the moveable access wall can be invaginated into the patient-care area, allowing the procedure to be carried out using two or more people on both sides of the bed. When the procedure is finished, the moveable curtain wall automatically returns to its resting position.

All supplies are passed into the unit through the pass-through located in the end wall. Doors of this pass-through are interlocked, so that only one can be opened at a time. Material is placed into the pass-through, the door is shut, gauntlets are donned, and hands and arms placed into the unit. The inner door of the pass-through is then opened and material taken out of the pass-through box and brought into the chamber. Dirty linen and trash are placed in plastic bags and removed from the chamber by placing them into a second plastic bag located outside of the unit but sealed to an aperture in the lower curtain wall. Bedpans are placed in plastic bags, the bags sealed, and then removed through the pass-through.

All patient-support equipment, such as intravenous bottles, respirators, monitors or pacemakers, are kept outside the unit itself. Only intravenous tubing, electrode leads, or air ducts go through the walls to the patient. In order to facilitate the entrance of patient leads, the overlap access panel walls are 12 inches from the back wall of the chamber and a solid panel closes the remaining space, allowing easy entrance for a variable number of tubes or conduits connecting the equipment outside the unit with the patient within. This arrangement allows the efficient use of nursing time, for intravenous bottles can be changed, drips regulated, respirators adjusted, or monitors observed as on the open wards. The unit as a whole provides a critically ill patient, requiring maximum clinical observation and minute to minute physiologic monitoring, with effective protection from cross-infection without requiring the attendance of a specially trained and individually assigned medical team.

Clinical experience

Over the past five months, 20 critically ill burned children have been cared for using the cluster of four units in the Shriners Burns Institute in Boston. The age range of these patients was from six months to fifteen years, and the proportion of body surface burned ranged from 30 to 80 per cent. During this initial period an intensive investigation has been carried out in order to determine the clinical efficiency of the unit in preventing forward or reverse cross-infection, either by the airborne or the contact route, as well as its efficiency in dealing with the problems of auto-infection. Second, the program has included investigations into the methods most efficient in the delivery of medical care.

These studies into the problems of patient care delivery, have demonstrated the feasibility of performing all but the most complicated surgical manoeuvers from without the unit by way of the access wall. The wall provides some increase in awkwardness and limitations in mobility, which decrease as individual experience is gained. However, for the usual types of nursing and medical care required for the critically ill, such as burn dressings, debridements, or even skin grafting under local anesthesia, the access walls are readily useable by nurses and doctors without any previous experience with the device. Preliminary evaluation of the studies being carried out to evaluate the human interaction with the mechanical device have demonstrated the need for adjustments, not only in the physical delivery of medical care when it is provided in a bacteria-free way, but also an adjustment of the psychological attitude toward the delivery of patient care by medical and nursing personnel. A physical barrier placed around the patient gives the impression of preventing direct patient contact. This impression leaves those providing medical care ill-at-ease, which progresses to anxiety as the clinical state of the patient becomes more critical. This is true both for nurses and doctors and, in our experience, it is no easier for a nurse to feed a patient through the access wall than it is for the physician-in-charge of respiratory care to manage a patient in respiratory failure when a patient is inside the unit and he and the respirator are outside. It is only through continued use and the comfort gained by experience that allows the medical-care staff to change their psychological as well as physical approach to isolation care with comfort.

The experimental program evaluating the efficiency of the bacteria-free patient-care unit as an isolating device has been more clear cut and encouraging. In the 20 patients studied thus far, there has not been a single case of cross-infection, either by the airborne or the contact route. Further, 11 of the patients placed in the unit have had serious burn wound infections at the time of their admission. All of these burn wounds contained two organisms and several had four organisms. In all of the patients observed, the number of bacteria species involved in the wound decreased and, in addition, a number of bacteria, as judged by the subjective criteria of plate growth (using the scale: abundant, moderate, scant and zero growth) or of wound cultures, have steadily decreased with the time the patient was in the bacteria-free environment. This tendency towards decreasing the concentration of bacteria on a previously heavily contaminated burn wound is demonstrated by the clinical course of a 12-year-old female who was admitted to the hospital three weeks following a 35 per cent full thickness body burn. On admission her wound contained abundant growth of *E. coli* and abundant growth of *Staphylococcus aureus.* Both *E. coli* and *Staphylococcus aureus* concentrations in the wound decreased steadily, and at three weeks following admission to the unit *Staphylococcus aureus* was no longer present and there was a scant growth of *E. coli.* The patient was then removed from the bacteria-free environment and five days later was shown to have wound contamination with Pseudomonas. The patient was placed back in the chamber with abundant Pseudomonas. Over the next month the concentration of bacteria in the burn wound steadily decreased until the pseudomonas disappeared from the burn wound four weeks after her return to the unit.

We have noted no incident of auto-infection of the burn wound with fecal flora while the patient has been housed in the bacteria-free environment. This observation is complicated by a clinical use of silver nitrate in our burn treatment, and therefore cannot be taken at its face value. The efficiency of bacterial removal by the air-handling system, however, has been shown to be very effective. In a study designed to quantify airborne viable particles and then to compare their concentration with the ward immediately outside the bacteria-free enclosure, it has been shown that the bacterial concentration of the ward during quiet activity has been less than 10 colony-forming units per cubic foot of air. In the unoccupied chamber, the bacterial count has been consistently zero organisms per cubic foot of air. Figure 4 compares the ward and chamber bacterial concentrations during periods of occupied quiet activity, dressing change, and bed-making. The figures show that a concentration of bacteria within the chamber remains at a near zero level through the purging effect of the air flow even during times of intense medical activity such as dressing change or bed-making. Through this method, the possibility of airborne auto-infection is greatly reduced.

Summary

A method for the delivery of bacteria-free patient care, effective both for forward and reverse isolation, has been described. Evidence indicates that the method also reduces the hazard of auto-infection while the patient is within the bacteria-free environment.

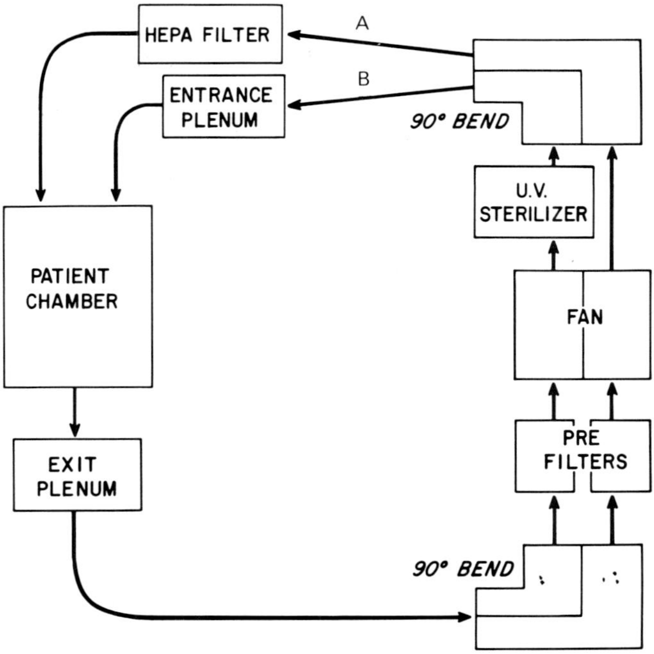

Figure 1. Block diagram of two mechanical air-handling and bacteria-removal systems using similar patient chambers and exit plenums.

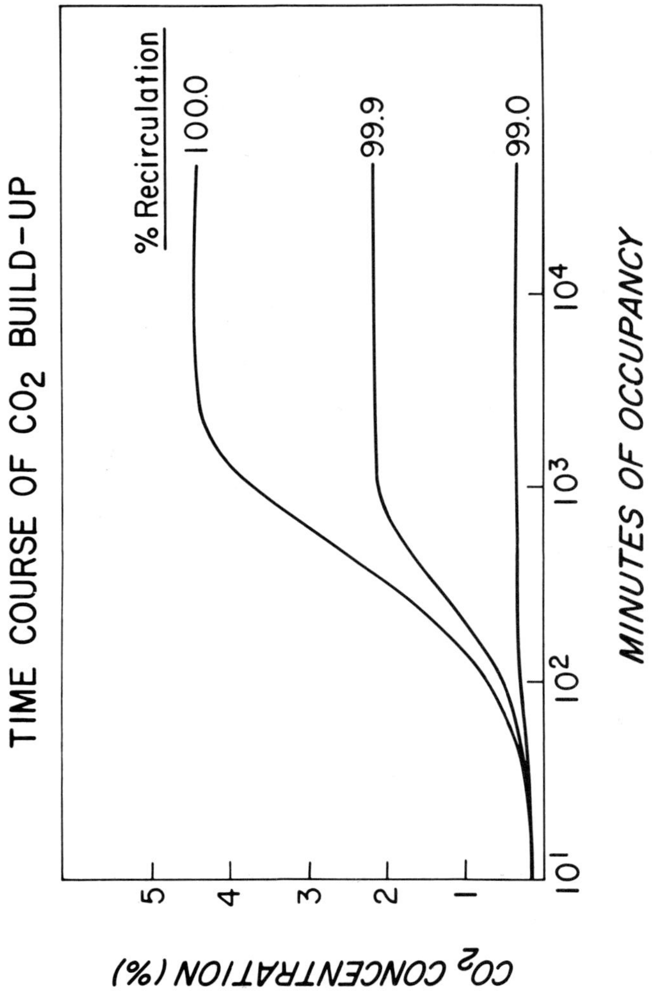

Figure 2. The time course of build-up of CO_2 in enclosed chambers with various percentages of air recirculation.

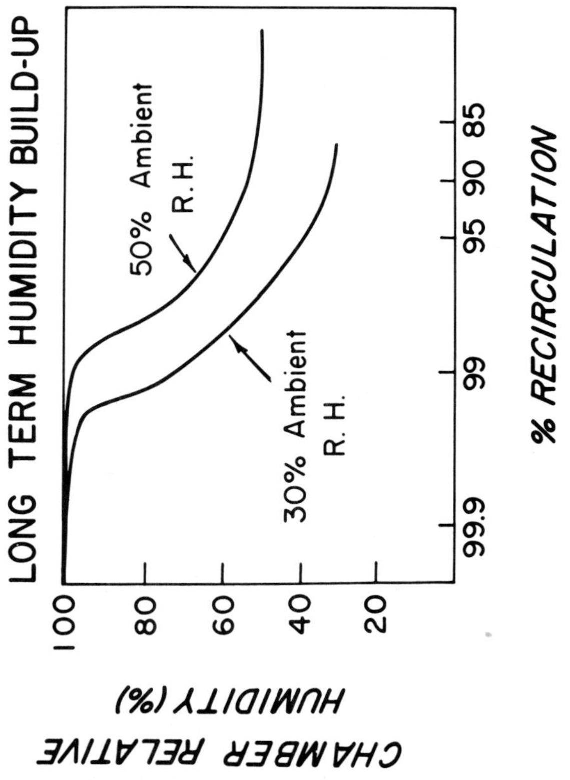

Figure 3. Long-term build-up of humidity in an enclosed chamber with various percentages of air recirculation during periods of 50 per cent and 30 per cent ambient relative humidity.

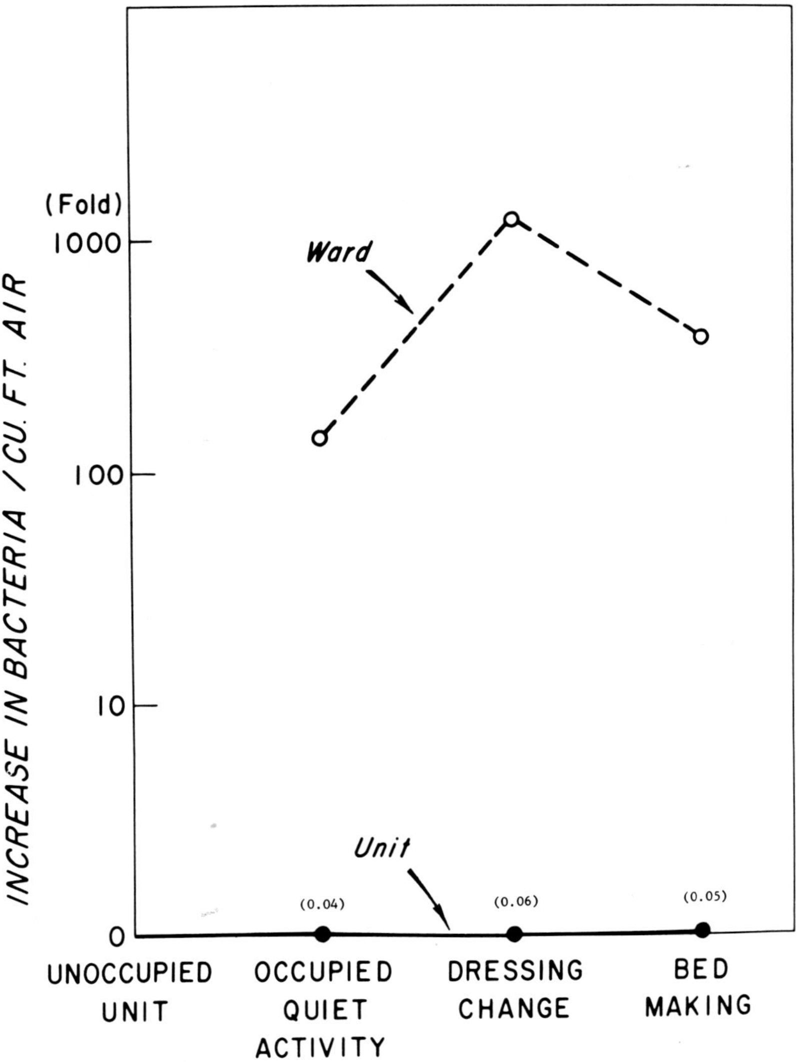

Figure 4. The efficiency of the bacteria-free nursing unit in reducing the bacterial content per cubic foot of air. The internal environment of the chamber is compared with the immediately adjacent ward.

The microflora of man in confined, controlled environments: review and analysis

F. B. Gordon

Naval Medical Research Institute, Bethesda, Md., U.S.A.

The insatiable curiosity of man combined with his modern engineering technology has led to beginning exploration of two new frontiers which, a short time ago, had been entered only in the imagination of science fiction writers. These are, of course, extraterrestrial space, and the depths of the seas.

The vehicles or habitations by which man is now entering these environments necessarily take the form of mechanical barriers, severely restricting the inhabitant to a small, completely enclosed space with some sort of life-support system which provides an artificial environment usually quite dissimilar to that at the surface of the earth. For explorations now under way or contemplated, extended periods of time spent in such environments are necessary.

The medical services responsible for protecting the health of the daring explorers who man these vehicles have a responsibility to determine the special hazards that may occur with long-term residence in such environments, and among the potential hazards is that of infectious disease.

This presentation will first examine briefly the nature of the medical microbiologist's task when he attempts to identify and assess any special infectious disease hazards that may be associated with lengthy residence in confined environments, especially those abnormal in gaseous content and pressure. Attention will then be directed to the rather numerous observations already made on the human indigenous microflora of persons in restricted, closed invironments, and finally an attempt will be made to draw conclusions from these studies, and to suggest approaches to the problems remaining.

A number of questions can be asked in this area of investigation, for which very few answers exist at present. One can ask, for example, whether a parabaric*

* 'Parabarosis' is newly coined to fill a definite need for a term to indicate departures from normal in a gaseous environment, either in the direction of greater or lesser pressure and/or qualitative differences from normal air, such as pressure of exotic gases, or abnormal ratios of ingredients of air. The adjective, 'parabaric', follows naturally.

environment will change the susceptibility of the vertebrate host to microbial infection. One can postulate changes in epithelial barriers due to direct contact with the abnormal atmosphere, or more deep-seated alterations in resistance factors. Various types of evidence can be cited showing the parabaric conditions will shift the host/microbe interaction in one direction or the other. The beneficial effect of hyperoxia on infections with obligate anaerobes has been demonstrated experimentally (1) and clinically (2). In other instances, one suspects that the effect is on the host. This is particularly apparent when one studies the effect of oxygen tension, with and without abnormal barometric pressures, on pulmonary infections. For example, increased O_2 tension is protective to mice with a chlamydial pneumonia, but renders an influenza viral pneumonia more severe (3). Several laboratories have reported increased severity of experimental viral or bacterial pneumonias in mice kept at simulated altitude but with a normal partial pressure of oxygen (4, 3).

Another question is whether the abnormal gaseous environments to be used (and in space, the factor of zero gravity) will change the properties of infectious aerosols, so that the incidence of interpersonal transmission of pathogenic micro-organisms will be significantly altered. Very little is yet known about the direct effect of parabarosis on microbial aerosols, a problem that should be relatively easily approached experimentally providing one can assemble the necessary hardware. Study of the effect of zero gravity will 'take a little longer'. Nevertheless, it can be assumed from what is already known concerning the effect of gravity upon deposition of infectious particles in the respiratory tract, that zero gravity will change the pattern of impingement normally seen. Gravity has been shown to be a significant factor in the deposition of particles of one to several microns in size in the periphery of the lung. In the absence of gravity fewer such particles will be deposited. This apparent advantage will probably be offset by the greater numbers of aerosolized particles of larger size—particles that would ordinarily be rapidly removed by sedimentation, but when inhaled will impinge on surfaces of the upper respiratory tract. It can be postulated, therefore, that in a space cabin the proportion of infectious particles contacting the upper respiratory tract, compared with those contacting the lower, will be greater than that seen on the surface of the earth.

Also a greater proportion of large particles in suspension in an infectious aerosol can be expected to influence the rate of loss of viability, other factors being equal.

One of the questions asked when this general problem was first enunciated was whether confinement in a restricted, altered environment will result in quantitative and qualitative changes in the normal indigenous microflora of man. Maintenance of the normal microflora of the alimentary canal and of the skin is regarded as an important factor in sustaining the resistance of these epithelial barriers to invasion of pathogens, and it is therefore important to avoid significant alterations in these microbial populations. Complicating the picture is change in standards of personal hygiene due to difficulties of bathing, hand-washing, tooth-brushing and the like, and an altered diet, such as that developed for astronauts, might be expected to result in changes in the microbial population of the gut. Additional factors to be considered are input of bacteria into the environment from imperfect methods of waste collection and disposal. Without the possibility of reduction of adverse or potentially dangerous elements in the atmosphere through the normal process of diffusion (and the lack of sedimentation in a zero-gravity situation), it is clear that a potential problem exists in the form of accumulation of micro-organisms in the environment as aerosols, or on interior surfaces from which they could become aerosolized or otherwise contaminate the crew, the food, water, etc.

It is apparent that potential microbiologic problems abound in restricted, abnormal environments, and because of the multiplicity of factors, the possibility of isolating any one or two for evaluation even in an experimental system is not good. Nevertheless, it is important to make whatever observations are feasible, and this brings us to the main portion of this report.

Over the past six years, a number of microbiologic studies have been made by the space agencies on human volunteers and their immediate surroundings in restricted or closed environments which more or less closely simulate the conditions found in space vehicles. Many of these experiments were conducted in air at ambient pressure; others were performed at simulated altitude with the gas phase similar to that of spacecraft. Studies of similar types have been performed by the U.S. Navy aboard submarines on prolonged submerged patrols, on volunteers during experimental occupancy of underground shelters, and in experimental diving chambers. In many instances the microbiologic studies were made only incidentally during the course of experiments that had other primary objectives, and the methods and scope of such studies have varied widely. It has been difficult to interpret the results, including the significance of observed changes, and comparisons of observations performed by different investigators are not easy. Nevertheless, an attempt was recently made to accumulate this information into one document (5), summarize it, and determine what significant conclusions can be drawn from it. The document cited is based on 26 studies (6) or series of studies of different types as tabulated in Table 1. In most of these studies various areas of the body of the personnel in the chamber were cultured regularly, some extensively, using as many as eight skin sites plus nose, eye, oral cavity, pharynx, and stool. In several the atmosphere was cultured quantitatively or semi-quantitatively, and in others, cultures were taken from interior surfaces of the chamber. In order to summarize, the results of culture are put into the five categories used in Table 1 with indicators of rise, fall, or a steady state of the microflora, high or low levels, or change, as interpreted in the footnote of the table.

There were essentially four types of atmosphere employed in the studies summarized:

1. Studies in air at ambient pressure. The studies performed in one or another sort of spacecraft simulator, in air at one atmosphere of pressure, are accumulated in the first section of the table (Codes A1, 5; A2, 3, 4, 6; C1, 3; B1, 4). In the first two

groups air was supplied as required and there was recycling through a CO_2 adsorber and dehumidifier, but a microbial filter was not placed in the line.

In the two others the air was recycled through filters, CO_2 scrubber, dehumidifier, etc. This type of atmosphere is represented also in the submarine environment, Code F1.

2. *Simulated altitude.* The second type of atmosphere (Codes A7, A8, A9, C2) was that resembling the gas phase of a space cabin in which the pressure was that of 17,000 feet altitude or higher, and O_2 was supplied in a partial pressure equal to or greater than that at sea level. Ordinarily the atmosphere was recycled for adsorption of carbon dioxide and was passed through filters. Study A9 was an exception in which a spacecraft atmosphere was maintained with only oxygen being introduced into the chamber and no recycling. The studies coded as D1, 2; E1 and E2, which were observations in actual spacecraft, grounded or in flight, also belong to this second category.

3. *Air, constant flow-through.* Observations coded as G1, 2 were conducted in an underground shelter in which fresh outside air was filtered and pumped constantly through the shelter, at a pressure only slightly greater than ambient.

4. *Hyperbarosis.* For the several studies coded as H1, 2 the experimental diving chamber was at simulated depths of 200–600 feet and the gas was a mixture of He with O_2 at approximately normal partial pressure.

Microflora of atmospheres

Let us first consider data on the microflora of the atmospheres used. Few tests were designed to provide direct comparisons between atmospheric microflora in normal and parabaric atmospheres, with other factors constant. In fact, any direct effect a spacecraft atmosphere has on microflora will be detected, it may be assumed, only by more sensitive types of observation than have been yet employed for this purpose. The experiments of study A7, however, give some basis for comparison, although only total numbers of bacteria recovered on exposed plates are reported. There were two subjects in each test, in a space-flight simulator of 380 cubic feet capacity. Oxygen concentrations varied from ambient to 100 per cent, depending upon the pressure. The results, with simulated altitudes and duration of observations, are given in Table 2. It may be seen that all counts, with a few exceptions, are in the same general range, although the counts made at sea level appear to be somewhat more irregular. It is not recorded whether any change in the life-support systems for these sea-level tests can account for their possible difference from the others. It is also apparent that there was no increase in atmospheric microflora over the period of the tests. Inspection of Table 1 shows that this was not always the case.

Of the nine series of studies in which the atmosphere was examined, two of them (A1, 5; A9) showed marked increases in bacterial 'load' in the atmosphere described as increasing from day 0 and later levelling off or dropping somewhat. The five experiments of A1, 5 were conducted in air at ambient pressure, which was recycled through a CO_2 absorber and an air conditioner, but *not* through a microbial filter. Figure 1 is a composite chart which shows the increase in atmospheric micro-organisms as determined by plates exposed to the air for fixed periods. The predominant type on the plates was recorded as staphylococci. The other study (A9), in which an increase occurred in the microflora of the atmosphere, was performed in a chamber at a simulated altitude of 27,000 feet, using 100 per cent O_2. Humidity and CO_2 were kept at desirable levels by a continuous flow through of O_2 with no recycling. The atmospheric bacteria, found in increased numbers during the course of the study, were described as 'largely staphylococci or micrococci, gram negative rods and to a lesser extent, streptococci'. In both of these studies an almost total absence of personal hygiene was achieved deliberately (no baths, restricted change of clothes, etc.) and it is not possible to evaluate the relative significance of this factor in comparison with lack of recycling with filtration in producing the increased level of microflora in the atmosphere. In contrast to these two reports, the atmospheric bacteria in seven other studies were reported as low or decreasing in four instances, and as steady or moderate in three. In all but one of these latter experiments recycling of atmosphere was occurring through systems that included a microbial filter. Table 2 gives data from this type of system (study A7) and Table 3 provides a sample of the results obtained with another such test (study C1). The figures indicate that the control systems used were adequate to provide a low level of atmospheric microflora.

One of the studies in which the microbial level of the atmosphere was reported as steady was the two underground shelter experiments (G1, 2) in which 100 men spent two weeks in a quonset hut that measured 48×25 feet on the floor. Although facilities for personal hygiene were only minimal no increase in atmospheric microflora was reported. This study used a flow-through type of ventilation using outside air, and is in this way similar to study A9, in which the atmospheric flora increased. The recorded data are insufficient to compare rates of flow.

There is a very clear indication in these studies that filtration through microbial-type filters of the atmosphere in this type of confined environment will control any reasonable bacterial load. More pertinent questions, perhaps, concern margins of safety, redundancy, or back-up systems in case of mechanical failure. Although these become important when one is in a situation where it is impossible to shut down and ring for a repairman, they are primarily engineering problems and solutions are attainable. Satisfactory microbiologic control of the atmosphere in spacecraft of the future seems to be a reasonable assumption. National Aeronautics and Space Administration personnel supplied the following information. In advance designs the spacecraft atmosphere will be subjected to three systems that will remove contaminants. A catalytic burner or oxydizer for removal of chemical contaminants will operate at a temperature of 700°F and cabin atmosphere will pass through at a flow rate of 3–6 CFM, giving a dwell-time of the gas inside the unit of about one second. Part of this system is a presorber of copper sulphate and a postsorber of lithium carbonate, both of which will reduce the bacterial count. Three blowers will operate at 1000 CFM as part of the system for temperature, humidity control, and circulation of the atmosphere. In the system designed primarily for bacterial control fans will operate at 750 CFM and be equipped with a 'roughing out' filter of 50 μ pore size, and a HEPA filter with $0 \cdot 3$ μ efficiency. This bacterial-control system was designed against a requirement of no more than 20 viable particles per cubic foot of gas. It is expected that final counts will be less than one particle per cubic foot. The essential features of this design have already been used in some of the tests summarized here.

Microflora of surfaces

No clear relationship is evident between microfloral counts on chamber surfaces, and data from other cultures. On the basis of the reports from C1, 3 and B1, 4 it may be seen that discrepancies between atmospheric levels and surface levels can occur. A potential source of micro-organisms in confined spaces is represented by the growth of bacteria and fungi in or upon materials that become damp. Presumably dampness will not occur in spacecraft so long as relative humidities remain under control at the levels planned. The dimensions of this problem in undersea shelters are somewhat greater.

Microflora of skin

A principal source of micro-organisms in the closed environments was, no doubt, from the skin of the personnel. Bacteria-laden particles are constantly being shed from the normal human skin at an estimated rate of many thousands per minute. Under the conditions of poor or absent skin hygiene in studies A1, 5 and A9 increases in skin microflora occurred and presumably greater numbers were shed into the environment. This, with absence of recycling through a microbial filter, resulted in greater atmospheric counts, as discussed in the preceding section. Figure 2 shows a composite estimate of numbers of bacteria in the axilla of the four men in an experiment of study A1, 5, covering a period of 42 days. Figure 3 gives similar figures for the groin.

In study A5, the wearing of space suits coincided with culturing of enterobacteria from the axilla; when the suits were no longer worn no more enterobacteria were cultured from that site. Some degree of maceration of the skin of the feet, with increased numbers of bacteria cultivated from the feet, has also been observed (C2).

Other evidence indicates that persons vary in the degree to which skin bacteria are shed. It was observed in study C3 that microbial counts of the air samples increased when one of the four operators who worked on shifts was present in the chamber.

The increase in skin count in experimental divers (Hl, 2) was directly related to immersion in contaminated water and need not be considered further here.

In spite of poor hygiene and increased bacterial counts of the skin in some instances, no significant skin infections were observed. Lack of a significant change in numbers of bacteria cultivable from the skin occurred in those tests in which a reasonable degree of personal hygiene was possible. As we have seen already, these were mainly the tests in which bacteria were filtered out of the air.

Microflora of oral cavity and pharynx

In most studies no remarkable changes were noted in numbers or types of micro-organisms in the mouth and throat. In contrast, in study A9 a definite increase in numbers of bacteria in mouth and throat were recorded. Streptococci accounted for most of the total increase, with some due to increases in staphylococci or micrococci. In this experiment two subjects were retained in a restricted environment in ambient air, for comparison with the six subjects who were at 5 psia in 100 per cent O_2. No significant differences between the two groups were seen in the bacteriologic observations.

Study C1 records first a decrease in normal microflora of throat and nose followed by replacement, in some cases almost completely, with *Staphylococcus aureus*, during the 30-day test. The precise numbers were not reported, but the ratio of the two types is shown in Figure 4. In the same study a marked increase was noted in numbers of bacteria referred to as organisms of Vincent's angina, as determined by direct smear of teeth, gingivae, and tonsillar areas. The increase was found in all subjects.

Microflora of feces

The changes observed in fecal flora can probably be related largely to changes in diet. Definite alterations in populations of micro-organisms occurred, but there were often wide variations between subjects in the same experiment. In study A5, for example, in one subject there was a shift in the *E. coli* population during the course of the test from 100 per cent non-typable strains to 50 per cent of the isolates being a typeable pathogenic type. This was attributed to an altered diet. It is not possible to detect effects of any of the other factors in the tests.

Interpersonal transmission of pathogens

In considering the transmission of pathogenic micro-organisms in these restricted environments we will exclude the two series (submarine studies, F1; underground shelter studies, G1, 2) in which more than a few subjects were involved. In both the studies mentioned febrile respiratory disease occurred which in some instances was due to interpersonal transmission within the environment under study. In the other experiments, in which there were never more than six subjects in the same chamber, little or no evidence for transmission of disease was seen—in fact illness was never a problem. Considerable attention was paid, however, to evidence for transmission of potential pathogens, such as staphylococci or enterobacteria, among the subjects.

Study A7, performed in a space-flight simulator, was specifically designed to determine the extent of transmission of potential pathogens. It consisted of 11 separate experiments lasting 14–30 days with two men each. They were confined in an elliptical cylinder of a total capacity of 380 cubic feet. The environment varied from ambient pressure to that of 33,500 feet, and the O_2 concentration ranged from ambient to 100 per cent. The results of cultures of the atmospheric microflora were discussed above (cf. Table 2). Observations were made of bacteriophage types of staphylococci in throat cultures, and of potential entero-pathogens in fecal cultures, in order to detect transmission of these types between the two subjects. There were one or possibly two clear instances, in a total of five reports, of transfer of staphylococcal type cultured from the throat. Among the enterobacteria used as tracers were *Proteus mirabilis*, *P. morgani*, *Salmonella arizona*, Providence group, *Aerobacter aerogenes*, *Alcaligenes*. In a few instances, in a total of 11 reports, convincing evidence of transmission was found. Examples of these are shown in Tables 4 and 5. The study depicted in Table 5 was extended beyond the time in the test chamber. Disappearance of the member of the Providence group from the stools of Subject 21 occurred very soon after termination of confinement. Subject 22 was regarded as the true carrier of the micro-organism. His stools were positive two months after the termination of the test, and members of his immediate family were also harboring the same type. Perhaps of greater significance is the failure to detect transfers between two subjects living in such crowded conditions. Examples of failure of transmission are seen in Tables 6 and 7. The authors of the report make the appropriate comment that other factors, in addition to exposure, determine whether a micro-organism will establish itself as a part of a person's indigenous microflora. In another study (A2–6), on micrococci only, transfer of phage-typable staphylococci among subjects and environments were apparently temporary only. During the 60-day test (C2) involving four men, two of them entered the chamber with *S. aureus* in their nose. It was never cultured from the other two crew members. In none of the studies was there any clinical illness related to the carriage or the transmission of potentially pathogenic bacteria.

DISCUSSION

At this point it is appropriate to ask what has been learned from the studies performed so far, only partially reviewed here. It is first desirable to recall that these studies show great heterogeneity in objectives, methods, detail of information reported, and probably in talent of the personnel responsible. Some experiments were designed primarily to test the effect of special diets, some to detect contamination from life-support and waste-management systems, and others had the more general objective of looking for any remarkable changes anywhere in the microbiologic picture. They varied greatly in the physical environments employed. It is not surprising, then, if it is difficult to draw definitive conclusions of a kind that pleases the experimentalist. With a few exceptions that are difficult to explain, it may be said that really nothing much did happen, as far as can be seen, to the micro-organisms or to the host-parasite relationship, under the conditions used and within the time of observation. This may be disconcerting to the researcher, but is reassuring to the physician responsible. It should be realized, however, that a series of negative results is never a safe base for extrapolation.

It is also appropriate to inquire whether the type of study under review here, i.e. the extensive programs of culturing of body surfaces and environment, can provide the answers needed to questions remaining. The studies reviewed represent tremendous investments in materials and man hours. For example, the single experiment of 34 days' duration, involving eight subjects, and coded as study A9, involved over 18,000 primary plate cultures. Over 150,000 secondary cultures were made and 10,000 microscopic slides were prepared and examined. In the series of experiments coded as A1, over 50,000 primary cultures were made and the number of secondary cultures approached 1,000,000. When one considers the paucity of definitive information that can be derived from the series of studies, the need for pause and reflection is obvious. In raising this question, we must not underestimate the value of the investigations made. They were designed primarily to be exploratory and orienting. But we must ask what should come next. Perhaps the state of the art has not reached the level required to make this full-scale type of approach worth while. Actually, our methods of quantitation of cultivable bacteria have advanced very little beyond that provided 90 years ago by the suggestion of Frau Hesse that agar be put into culture media. New approaches to the problems of microbiologic surveillance and monitoring are much needed. In the meantime, the answer to the dilemma possibly lies in greater selectivity, with more intensive and careful investigation in circumscribed areas.

You will note that most of the investigations performed so far in this general area have been directed primarily to space flight. Although the parabaric conditions of deep and prolonged dives in the sea, i.e. helium–oxygen mixtures with increased pressures, should be considered as part of this general problem, little can as yet be said about the effect of these conditions of infections or infectious agents. No remarkable results were encountered in the limited study coded as H1, 2. Other research is under way in this area, however, and it has already been shown that *S. aureus* growing under pressure of 1000 psi of helium, or of helium plus O_2 at normal tension, exhibited a decreased susceptibility to penicillin (7).

A potential hazard to men in parabaric environments is represented by altered O_2 pressures. Present procedures for both astronauts and aquanauts, especially when beginning or ending their missions, include exposure to O_2 tensions somewhat above normal. The dramatic effects of O_2 toxicity are well known and can be avoided, but little is known about more subtle effects. One was cited earlier, i.e. the greater susceptibility of mice to pulmonary infection with influenza virus when exposed to increased O_2 tension (3). The level of O_2 (77 per cent at ambient pressure) was compatible with apparently normal life for at least many weeks, but mortality from influenzal pneumonia was significantly greater under those conditions. And it is disconcerting to find, as cited earlier, that similarly infected mice die sooner when in a spacecraft atmosphere (70 per cent O_2 in N_2 at 5 psia) designed to provide a normal alveolar O_2 tension at altitude (3, 4). Perhaps the percentage of O_2 in the atmosphere is significant, as well as the partial pressure. It is known that O_2 tension influences the outcome of viral-cell interaction and interferon may be a factor in this effect (8). Variation of O_2 tension was found to affect the level of interferon extractable from the lung after induction (9). The effect of O_2 tension on radiation damage to cells is well documented and damage to chromosomes by O_2 at increased pressure has been reported (10). The possibility of parabaric conditions affecting latent viral infections or other intracellular disease agents demands attention.

In a few of the studies summarized, a reduction in numbers of indigenous micro-organisms, or a 'simplification' of microflora, were reported. The evidence for these conclusions was not always provided and in other seemingly similar experiments such changes were not observed. Evaluation of these findings must be deferred. However, it has been postulated (11, 12), mainly on the basis of observations on animals kept in restricted environments, that prolonged isolation of minimal numbers of persons in closed environments, with sterile air, food, and water, and without 'reinoculation' by contact with the micro-organisms of numbers of people, will result in profound changes in the 'normal' microflora. Reduction in numbers and types of indigenous micro-organisms has been predicted, as well as shifts in the usual microbial populations, for example, from one level of the alimentary tract to another. This problem is related to that of isolation from currently circulating pathogens such as respiratory tract viruses. Astronauts returning from a journey of many months would be expected to show the type of susceptibility to respiratory tract infection seen in isolated Arctic communities, for example, with their first contact with the outside world in the Spring. The extreme of this effect is seen in the 'bacterial shock' seen in gnotobiotic animals after contact with conventional animals of the same species, and emphasizes the advantage of maintenance of the usual indigenous microflora. Further assessment of this aspect of prolonged isolation is indicated.

Perhaps at this time, it may be tentatively concluded that nothing immediate or dramatic is likely to happen, microbiologically speaking, as a result of the restricted environment and the abnormalities of atmospheric pressure and content encountered in space flight. As data accumulate from undersea investigations, more definite statements can be made concerning the type of parabarosis. The possible dangers of zero gravity, such as an altered pattern of deposition of aerosolized particles in the respiratory tract, can be approached only theoretically or indirectly at present. When it becomes possible to test such an effect directly this is a problem that should be on the agenda. Significant reduction or alterations in indigenous microflora must be watched for, but this can presumably be accomplished by judicious selection of time intervals, sites to be cultured, and types of micro-organisms to be assayed.

I wish to suggest, however, that potential hazards of a serious nature may more likely rest in areas of inquiry largely untouched as yet. The subtle effects that alterations in the atmosphere, especially in O_2 tension, can have on vertebrate and other cells must be examined more closely with respect to infection. The evidence so far available from experimental animals indicates that lung tissue, when directly exposed to altered atmospheres, does indeed behave differently than usual in the presence of infection. It is important to determine the extent and the mechanisms of such alterations in host–parasite interaction at the cellular level. This type of problem also has the merit of being approachable with present technics.

SUMMARY

The problems of the medical microbiologist in studying the hazards of space flight and prolonged dives in the sea have been briefly analyzed. Significant results from a series of studies on the microflora of man is confined and controlled environments have been summarized and some of the highlights of the studies presented. An attempt was made to draw conclusions from the summarized studies and to delineate remaining problems in this field of investigation.

REFERENCEES

1. Kaye, D. (1967). Effect of hyperbaric oxygen on clostridia in vitro and in vivo, *Proc. Soc. Exp. Biol. and Med.*, **124,** 360–366.
2. Brummelkamp, W. H. (1964). Reflections of hyperbaric oxygen therapy at 3 atmospheres absolute for *Clostridium welchii* infections, *Hyperbaric Oxygenation, Proc. 2nd Int. Cong.*, pp. 239–250, I. M. Ledingham, Ed., Edinburgh: Liverston 1965.
3. Gordon, F. B, and Gillmore, J. D. (1969). Unpublished observations.
4. Ehrlich, R. and Mieszkuc, B. J. (1969). Resistance to experimental bacterial pneumonia and influenzal infection in space cabin environment, *Aerospace Med.*, **40,** 176–179.
5. Vargosko, A. J., Gordon, F. B. and Wilkins, J. R. (1969). Microbiologic studies on man in closed environmental systems: Summary and interpretation of reported observations. (Report prepared for National Academy of Sciences, Space Science Board.)
6. References to the studies summarized in Table 1 and text are placed at the end of this list.
7. Schlamm, N. A., Coykendall, A. L. and Myers, C. E. (1968). Effect of penicillin in *Staphylococcus aureus* cultivated at high atmospheric pressure, *J. Bacteriol.*, **98,** 327–328.
8. Isaacs, A., Porterfield, J. S. and Baron, S. (1961). The influence of oxygenation on virus growth. II. Effect on the antiviral action of interferon, *Virology,* **14,** 450–455.

9. Huang, K-Y. and Gordon, F. B. (1968). Production of interferon in mice: Effect of altered gaseous environments, *Applied Microbiol.*, **16,** 1551–1556.
10. Moutschen-Dahmen, M., Moutschen, J. and Ehrenberg, L (1959). Chromosome disturbances and mutation produced in plant seeds by oxygen at high pressures, *Hereditas,* **45,** 230–244.
11. Luckey, T. D. (1966). Potential microbic shock in manned aerospace systems. *Aerospace Med.*, **37,** 1223–1228.
12. Bengson, M. H. and Thomas, F. W. (1965). Controlling the hazards of biological and particulate contamination within manned spacecraft. *J. Am. Assoc. Contam. Control*, **4,** 9–12.

References to coded studies listed in Table 1:

A1. Riely, P. E., Gieb, D. and Shorenstein, D. (1966). Determination of the indigenous microflora of men in controlled environments. AMRL-TR-66-33 (Aerospace Medical Research Laboratories, Wright-Patterson Air Force Base, Ohio).
A2. Lotter, L. P. *et al.* (1967). The potential hazard of staphylococci and micrococci to human subjects in a life support systems evaluator and on a diet of precooked freeze dehydrated foods. AMRL-TR-67-18 (see A1).
A3. Lotter, L. P. and Horstman, B. S. (1967). The potential hazard of staphylococii and micrococci to human subjects in a life support systems evaluator with elevated cabin temperature AMRL-TR-67-43 (see A1).
A4. Lotter, L. P. *et al.* (1967). The potential hazard of staphylococci and micrococci to human subjects in a life support systems evaluator and on a diet of liquid foods, AMRL-TR-67-21 (see A1).
A5. Riely, P. E. and Shorenstein, D. J. (1966). Microbiological flora of human subjects under simulated space environments. AMRL-TR-66-171 (see A1).
A6. Lotter, L. P. and Horstman, B. S. (1967). The potential hazard of staphylococci and micrococci to human subjects in a life support systems evaluator while on a simulated GT-7 mission. AMRL-TR-67-45 (see A1). Lotter *et al.* (1968). Distribution of straphylococci and micrococci among humans and their physical environment during simulated aerospace confinement, *Am. J. clin. Pathol.,* **49,** 414–423.
A7. Moyer, J. E. and Lewis, Y. Z. (1964). Microbiologic studies of the two-man space cabin simulator; interchange of oral and intestinal bacteria. SAM-TDR-64-3 (School of Aerospace Medicine, Brooks AFB, San Antonio, Texas).
A8. (*a*) Cardaro, J. T. *et al.* (1966). Study of man during a 56 day exposure to an oxygen–helium atmosphere at 258 mm Hg total pressure. X. Enteric microbial flora, *Aerospace Med.,* **37,** 594–596.
(*b*) Moyer, J. E. *et al.* (1966). Study of man during a 56-day exposure to an oxygen-helium atmosphere at 258 mm. Hg total pressure. XI. Oral, cutaneous, and aerosol bacteriologic evaluation, 597–600 (Follows (*a*), above.)

(*c*) Zeft, H. J. *et al.* (1966). Study of man during a 56-day exposure to an oxygen-helium atmosphere at 258 mm Hg total pressure. XII. Clinical observations, 601–604. (Follows (*b*) above.)

A9. Coburn, K. R. (1967). A report of the physiological, psychological, and bacteriological aspects of 20 days in full pressure suits, 20 days at 27,000 feet on 100% oxygen, and 34 days of confinement. NASA Contractor Report; NASA CR-708.

B1. Wilkins, J. R. (1967). 7-day closed-door test on the Langley integrated life support system: Air, water, and surface bacteriologic investigations. Langley working paper; LWP-431. Langley Research Center, Langley Station, Hampton, Virginia.

B2. Wilkins, J. R. (1968). Mircrobiologic studies on life support sub-systems for manned space flight. Presented at American Industrial Hygiene Assoc., St. Louis, May 1968.

C1. The Boeing Co. (1964). Manned environmental system assessment. NASA Contractor Report; NASA CR-134.

C2. McDonnell Douglas Astronautics Co. (1968) Sixty-day manned test of a regenerative life support system with oxygen and water recovery. Part II. Aerospace medicine and man machine test results. NASA Contractor Report; NASA CR-98500.

C3. Wilkins, J. R. (1969). Personal communication concerning 28-day evaluation of manned Integrated Life Support System. Langley Research Center, Langley Station, Hampton, Virginia.

D1. Borchardt, K. A. *et al.* [1968). A partial bacteriological profile of test pilots and their Apollo space vehicle during a simulated fourteen-day lunar flight, *Aerospace Med.,* **39,** 166–171.

D2. Medical Research and Operations Directorate (1968). The 2 TV-1 test medical report. NASA, Manned Spacecraft Center, Houston, Texas.

E1. Hotchin, J. *et al.* (1967). The microbiological flora of the Gemini IX Spacecraft before and after flight. NASA Contractor Report: NASA CR-972.

E2. Berry, C. A. (1969). Preliminary clinical report of the medical aspects of Apollos VII and VIII, *Aerospace Med.,* **40,** 245–254.

F1. Watkins, H. M. S. *et al.* (1967). Epidemiological studies aboard Polaris submarines. Report presented to Armed Forces Epidemiological Board Commission on Acute Respiratory Diseases, March 1967.

G1. Ramskill, E. A. *et al.* (1962). Studies of the Bureau of Yards and Docks Protective Shelter. I. Winter Trials. U.S. Naval Research Laboratory, Washington, D.C.

G2. Bogardus, H. F.(1968). Studies of the Naval Facilities Engineering Command Protective Shelter, II. Summer Trials, U.S. Naval Research Laboratory, Washington, D.C.

H1, 2. Schlamm, N. A. Personal communication.

TABLE 1

Summary of studies analyzed

Code[a]	Type of chamber	Atmosphere	No. tests	No. subjects	Duration, days	Hygiene	Results of culture: Atmos.	Surfaces	Skin	Oral cav. and pharynx	Feces
A1, 5	Life-support systems evaluator	Air	8	4	21–60	Poor	↑		↑	—	Ch
A2–6[b]	Same	Air	6	4	21–60	Poor		—	—	—	—
C1, 3	Space-cabin simulator	Air recycled	2	5, (4)	30, 28	Good	— (low)	High		Ch	
B1, 4	Same, with life-support systems (unmanned)	Air recycled	2	0	7, 17		↓	—			
A7	Space-cabin simulator	Spacecraft, recycling	11	2	14–30	?	—		—	—	Ch
A8	Internal-environment simulator	Spacecraft, recycling	1	6	68	Good	—		—	—	Ch
A9	Spacecraft simulator	Spacecraft not recycled	1	8	34	Poor	↑	— (high)	↑	↑	Ch
C2	Space-cabin simulator with life-support systems	Spacecraft, recycling	1	4	60		— (low)		—	—	

TABLE 1 *(continued)*

Code	Type of chamber	Atmosphere	No. tests	No. subjects	Duration, days	Hygiene	Results of culture				
							Atmos.	*Surfaces*	*Skin*	*Oral cav. and pharynx*	*Feces*
D1, 2	Apollo spacecraft, grounded	Spacecraft, recycling	2	3	8, 14	?		—		—	Ch
E1	Gemini IX	Spacecraft	1	2	3			—			
E2	Apollo VII, VIII	Spacecraft	2	3	6, 11			'Simplification of microflora'			
F1	Submarine patrol	Air recycled	8	100+	Av. = 56	Good	— (low)	— (low)	—	—	—
G1, 2	Underground shelter	Air	2	100	14	Minimal	—			—	—
H1, 2	Exp. diving chamber	Hyperbaric O_2–He	2	4, 5	4, 12	Minimal			↑ [c]	—	—

[a] Code numbers correspond to those of basic document (5).
[b] Observations on micrococci only.
[c] Related to contact with contaminated water.
↑↓ = Increase or decrease in bacterial counts.
— = Incidence steady; no significant variations.
Ch = Change in incidence of types.
Blank space—Not done or unknown.

TABLE 2

Bacterial aerosol counts of space-cabin simulator (taken from table 6 of study A7)

Day of flight	*Sea level, 17 days*		*27,000 feet, 14 days*			*33,500 feet, 17 days*		
1	51	16	33	32				57
2		26			31	60	34	
3	182		19	71	10		86	
4	57	53				80		
5			19	78				
6		104		40	66		50	59
7				93		85		18
8	102		30				45	
9						116		11
10	41			82	9			
11		115		78			112	
12			29			54		
13		TNTC		95	33			
14						36		36
15						38	55	
16		128				85		31
17	107	10					34	

Each count represents an average of five sampling stations (90-minute exposure/plate). Each column represents one test.

TABLE 3

Results of bacterial air sampling on day 20 of space-cabin simulator test (taken from from table 6 of study C1)

Sampling port no.	1	3	4	5	6	7	10	11
Total cubic feet sampled	15	20	20	30	30	30	2	30
No. bacteria recovered	16	3	1	2	5	1	106	0
No. bacteria per cubic foot air	1	<1	<1	<1	<1	<1	53	0

Port 10: Source includes waste reactor.

TABLE 4

Staphylococcal types from the two crew members during an 18-day test in a space-cabin simulator (adapted from table 1 of study A7)

Date	Phage pattern	
	Subject 13	Subject 14
4/10	—	7:47*
11	—	7
12	—	—
13	—	7
17	—	47
18	7	47
20†	—	7
21	—	—
22	—	—
23	—	—
24	7	—
25	—	—
26‡	7	7
27	—	—
29	7	—
5/2	—	—
3†	—	7
4	—	47
5	—	7
6	—	47
7	47	7
8‡	—	7
9	—	47
10	—	47

* In all instances 47 = type 47/54/75/83.
†, ‡ Entry, exit from space simulator (Tables 4, 5, 6, and 7).

TABLE 5

Enteric marker bacteria isolated from stool cultures of the two crew members during a 30-day test in a space-cabin simulator (adapted from table 4 of study A7)

Date	*Subject 21*	*Subject 22*	
3/9	Proteus	Prov.*	
12	—	Prov.	Prov.
15 †	Aerobacter	Prov.	
16	—		
17	—	Prov.	
20	—		
21	Prov.	Prov.	
23	Prov.	—	
24	Prov.	Prov.	
25	Prov.		
26	—		
27	—		
29	Prov.		
30		Prov.	
31		Prov.	
4/3	Prov.		
4	—	Prov.	
5	—		
6	Prov.		
9	—	Prov.	
11	—		
13 †	Prov.	Prov.	
14	Prov.	Prov.	
16	—	Prov.	
17		Prov.	
20		Prov.	
21	—	Prov.	
5/15	—		
22	—	Prov.	
28	—		
6/1	Proteus		
8		Prov.	
16		Prov.	
22	—	Prov.	

* Prov. = Providence type.

TABLE 6

Absence of interpersonal transfer of throat flora (adapted from table 3 of study A7)

Date	Subject 21	Subject 22
3/2	—	—
3	—	—
6	52*: *A. aerogenes*	—
8	*A. aerogenes*	83
13	*A. aerogenes*	—
15 †	52: *A. aerogenes*	—
16	—	—
17	*A. aerogenes*	83
20	*A. aerogenes*	—
22	*A. aerogenes*	83
23	*A. aerogenes*	83
24	52: *A. aerogenes*	—
28	*A. aerogenes*	83
30	—	—
4/4	—	—
6	*A. aerogenes*	—
9	52	—
11	—	—
12	*A. aerogenes*	—
13 ‡	—	—
19	52: *A. aerogenes*	—

* Numbers indicate phage types of *S. aureus*: 52 = 52/52A/80/81.

TABLE 7

Absence of interpersonal transfer of enteric flora (adapted from table 5 of study A7)

Date	*Subject 11*	*Subject 12*
2/14	Providence	—
15	*A. aerogenes*	—
16	*A. aerogenes*	—
17	*A. aerogenes*	—
18	*A. aerogenes*	—
19	—	—
20	Providence	
21	Providence	—
25	—	—
27[†]		—
28	Providence	
3/2	Providence	—
3	Providence	—
7	Providence	
8		—
12		—
14[‡]	Providence	Pseudomonas
18		Pseudomonas
19	Providence	—
20	Providence	
21		—
22	*A. aerogenes*	—

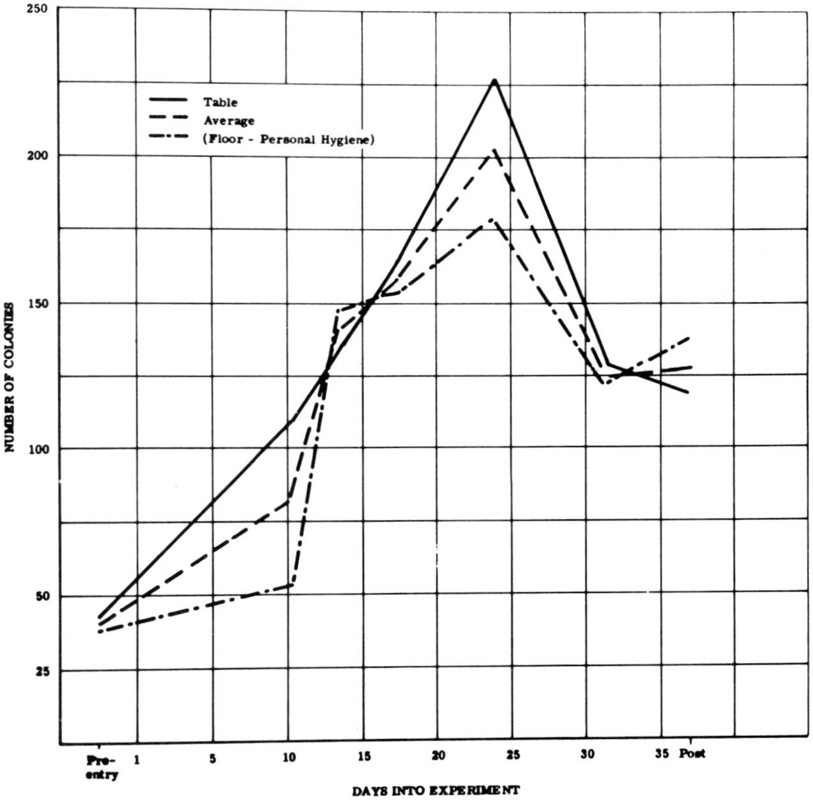

Figure 1. Composite graph of colony counts on culture plates exposed at various sites to atmosphere for 30 minutes. (Fig. 13 of study A5.)

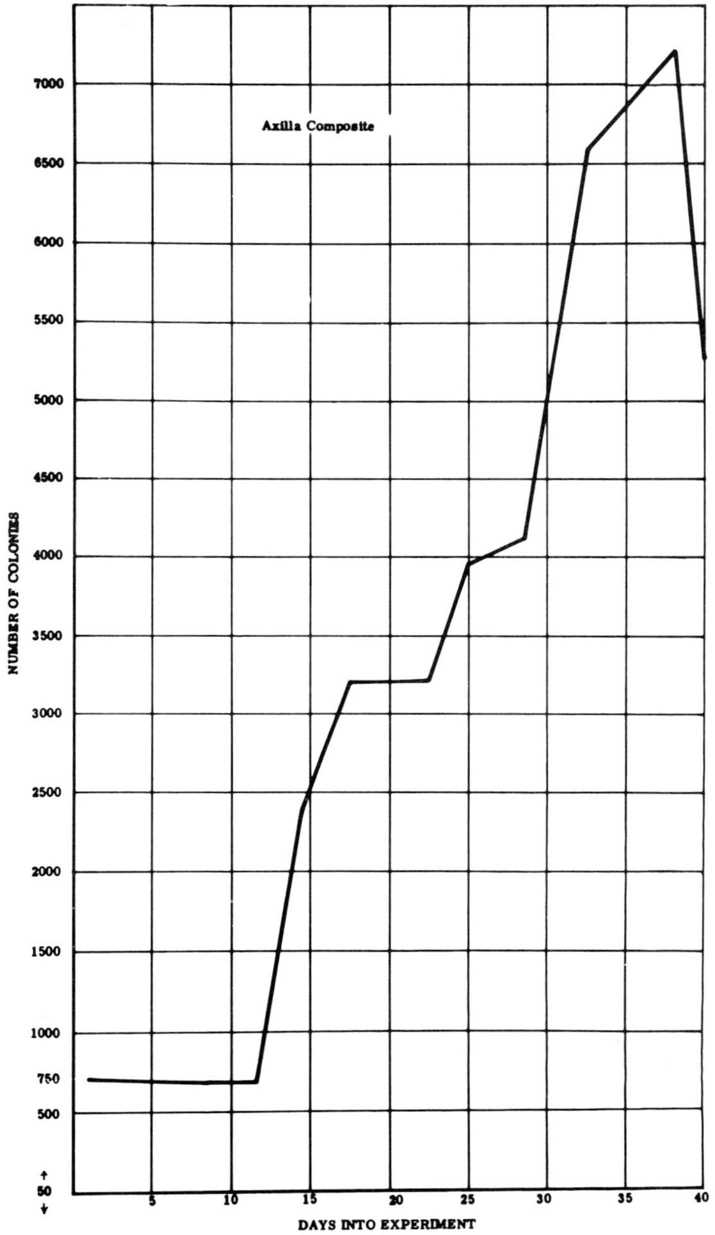

Figure 2. Numbers of bacteria cultured from axilla of subjects in Life Support Systems Evaluator; composite figures. (Fig. 3 of study A5.)

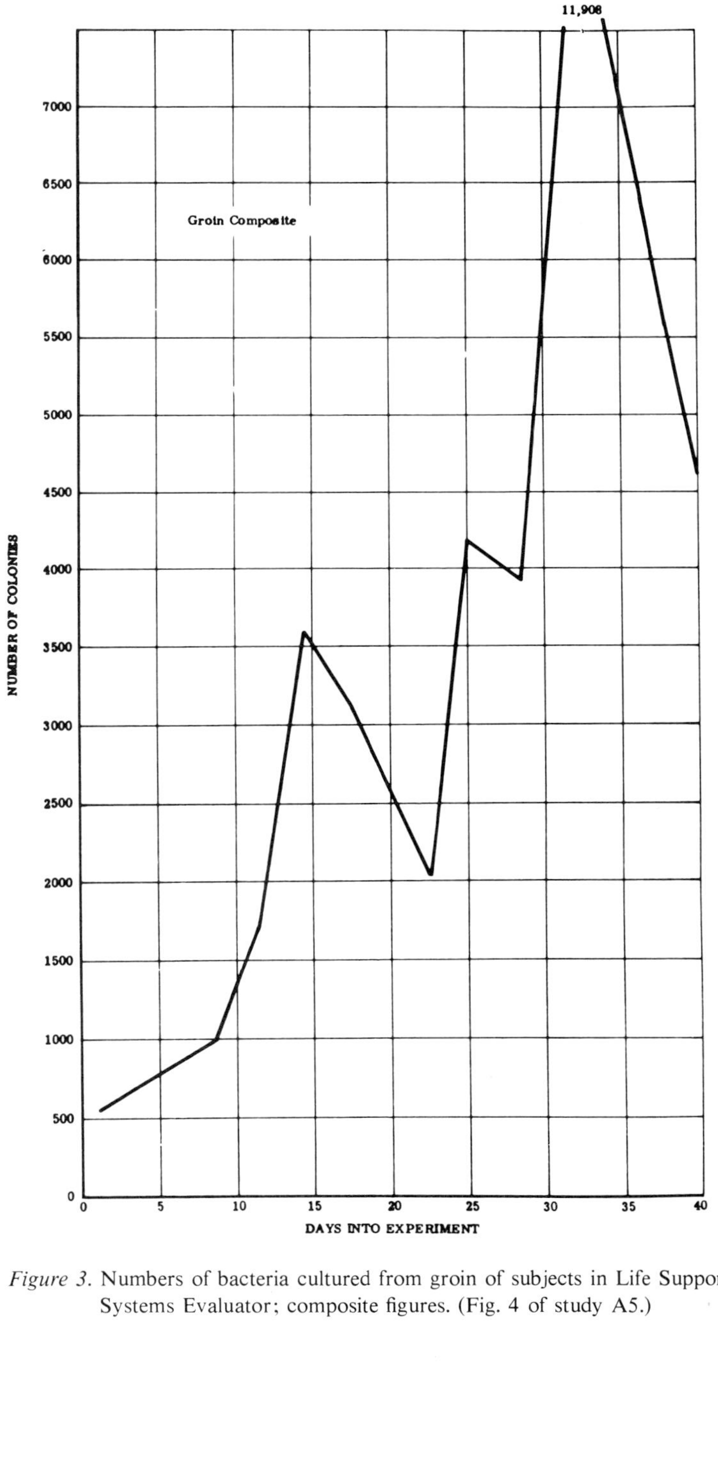

Figure 3. Numbers of bacteria cultured from groin of subjects in Life Support Systems Evaluator; composite figures. (Fig. 4 of study A5.)

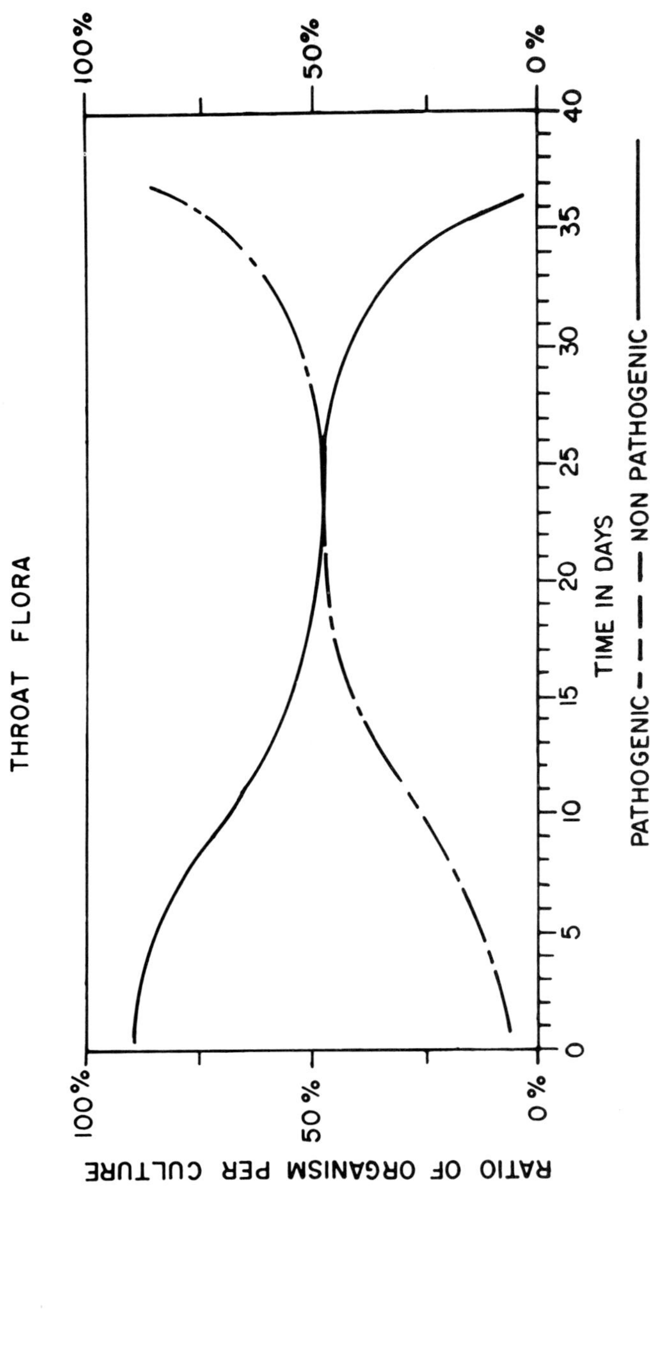

Figure 4. Relative numbers of pathogenic (*S. aureus*) and non-pathogenic bacteria in throats of subjects during test in space-cabin simulator. (Fig. 64 of study C1.)

Experience with germ-free environments in surgery in relation to design

John Charnley, D.Sc., F.R.C.S.

Wrightington

The performance of surgical operations in a sterile environment raises two questions: first, is it a practical proposition from the point of view of comfort and convenience for the surgeon and the economics of the hospital? and second, does it significantly reduce post-operative wound infection? I feel competent to discuss the feasibility and economics of operating in a sterile environment, but it is still too early, since the introduction of our latest aseptic developments, for me to indicate precisely what contribution to our reduced infection rate comes from the sterile environment and what from some other elements in surgical technique.

Sterile environments in the operating room

During the last 10 years my ideas on the design of an operating room for operations demanding extreme asepsis have gone through a number of stages of development, and I think it would be helpful to trace these steps historically.

From the very start of this work in 1958 I have insisted on isolating the sterile operating team from the rest of the personnel in the theatre by the use of some form of enclosure.

In 1958 the nasopharynx was usually regarded as the most likely source of pathogenic staphlococci, and I introduced vacuum-extraction masks to carry away expired air and keep the textile of the masks dry. We have retained this feature as much for the comfort it brings to the surgeon as for its bacteriological effectiveness. By extracting humid exhaled air and carbon dioxide from the dead space of the mask the surgeon is spared fatigue, and sweating is reduced which in itself is favourable to reducing infection rate as by glove puncture.

The method of introducing filtered air into the operating enclosure was dominated at first by the idea of 'downward displacement', which was a current phrase at that time. To avoid the turbulence inherent in introducing 1000 c.f.m. into an operating enclosure of only 350 cubic feet capacity I brought the air into the roof of the enclosure through three long cambric sacs. This fine cotton textile offered slight resistance to the air flow which distended the sacs and allowed the entering air to leak into the roof of the enclosure without the slightest turbulence. This technique of using fine cambric tubes or sacs in the roof is still a very valuable method of upgrading an existing theatre, because any volume of filtered air can be introduced without turbulence and with practically no alteration to the internal structure of an existing theatre (Figure 1).

It soon became apparent, as a result of settle-plate experiments and smoke tests, that if the incoming air was at the desired temperature of 65°F it interacted strongly with hot air rising from the bodies of the four persons inside the enclosure. The hot air generated by the surgeons proved to be much more significant than had been expected. The cool air descending in the centre of the roof of the enclosure enhanced the tendency of warm air to rise on the periphery of the enclosure and so entrain with the descending cool air. This flow-pattern was assisted also by the walls of the enclosure, which, being made in part of aluminium, picked up heat from the surrounding operating theatre at 75°F.

From these experiences it became clear that even changing the air in the enclosure about 100 times an hour, with the escape of air arranged at the foot of the enclosure, it was impossible to achieve the concept of 'downward displacement'. It is possible that we might have achieved this by heating the incoming air to 75°F, but with the surgeons' comfort in mind this was a step which I refused to consider.

One of my experiments in an effort to assist downward displacement was to add an air-extraction system in the first installation at the foot of the walls of the enclosure. This proved quite futile. To preserve positive pressure inside the enclosure it was not reasonable to extract more than about 300 c.f.m. with an input of 1000 c.f.m., and the disappointing feature of air extraction was the negligible effect it had on the movement of air more than half an inch from the apertures of the suction manifold. It would appear that sucking and blowing, aerodynamically speaking, are two very different things!

The next step was to plan for the entry of a larger volume of filtered air into an enclosure of the same ground plan area (50 square feet), in the hope of getting a downward draught of air at 65°F sufficiently positive to overcome the tendency to upward currents of warm air generated by the four persons of the operating team and the warm walls of the enclosure. This was a natural step towards the concept of 'laminar flow' which theoretically would sweep downwards all infected dust particles as soon as they were emitted from the gowns of the operating team and without turbulent mixing inside the enclosure.

A new operating enclosure was then designed and built (by the Manchester Regional Hospital Board with a grant from the Medical Research Council), and came into full operation in June 1966. The operating enclosure, 7 × 7 feet in ground plan area, was made on half-inch armour plate glass (to reduce heat transfer from the surrounding warm room and facilitate viewing by spectators) and it extended from the floor to the ceiling, a distance of 10 feet. The ceiling of the enclosure was replaced by a rectangular aperture, $5\frac{1}{2} \times 5\frac{1}{2}$ feet, communicating with a chamber of the roof, approximately 10 × 10 × 10 feet, and in the side of this the bank of pocket filters was housed.

By the added height of the chamber above and the enclosure reaching the ceiling of the operating room it was hoped that the flow of 4000 c.f.m. of filtered air would have time to 'sort itself out' into true laminar flow before reaching the operation site. When we came to carry out smoke tests at first it was disappointing to find that there was very violent turbulence inside the enclosure when an object was present simulating the patient on an operating table. Even though the linear air speed was not quite 100 feet per minute this was sufficient to cause strong upward currents when the downward air stream struck the body of the patient, and the surrounding surgical team. In other words I believe that the concept of 'laminar flow' in operating rooms is a fallacy, because it would only be achieved by aerodynamic streamlining of the bodies of the operating team and the patient.

None the less this installation has proved an enormous success, and it has pointed the way to what I believe will be the final perfect solution. The success of this enclosure is attributable to the very considerable dilution produced by large volumes of filtered air. We have reduced the original turbulence by introducing vanes, or louvres, in the aperture in the ceiling, which have been inclined in various directions, as a result of smoke tests, to direct a portion of the incoming air from the centre of the descending column to scour the inside of the walls of the enclosure.

In this enclosure, with 300 air changes per hour, we have regularly exposed blood-agar settle plates during the whole period of the operation (one to one and a half hours) and only rarely collected one colony. A particularly attractive feature of this installation has been the visibility afforded to visitors and post-graduates studying surgical technique. It is possible for six or more visitors to get a perfect view of the operating field without any impediment to the surgical team or any prejudice to the welfare of the patient.

Air filters

It has always seemed important to me to design for large volumes of filtered air, low consumption of power in fans, and long life of filters. With economic matters realistically in mind I have always held that it is unnecessary to have air filtration more perfect than 1–2 microns because dust particles which settle from air in slow motion are now generally recognized as being of the order of 10 microns or more. It would therefore be more accurate to describe my approach as being towards a dust-free environment rather than a truly sterile-air environment.

In large operating-room installations where there is adequate space available in adjoining rooms, I believe there is a lot to recommend large-size cloth filters of the pocket type in preference to the disposable paper-type filters that are used for absolute filtration. The cloth pocket filters can be laundered and used several times, but the changing of paper filters can be very expensive over the years. In our present installation, with a very simple gamgee pre-filter for 10–12 microns we have pocket filters still in action after nearly three years of continuous service. There has been only a slight reduction in their through-put of air as measured directly by air speed with a rotating vane anemometer built into the unit. It is true that we are in fairly clean air and partly in the country. The only reason why we shall eventually change these filters is the possibility of oxidization causing the fibre to perish. This installation has been in continuous use day and night passing 4000 c.f.m. during operating sessions, and 2000 c.f.m. during the rest of the week.

Humidity

The ideal operating enclosure would be provided with an efficent humidifying system after the air heater. Tissues exposed for a long time to rapidly changing air can be desiccated on the surface, and theoretically their resistance to infection might be impaired.

Illumination

Illumination is the great problem in designing an operating enclosure if one adheres to the erroneous concept of 'laminar flow'. It is difficult to get good vertical illumination without obstructing a downward flow of air in the centre of an enclosure. In Johannesburg, Professor du Toit is proposing to overcome this defect in a new installation by the use of sterile fibre-optics, light guides, which would seem to offer an excellent solution to some of these problems.

Future development in enclosure design

The present enclosure we believe has reduced our infection rate from the level of 5 or 3 per cent which obtained in the prototypes to 1 per cent. Despite the fact that for all intents and purposes we have sterile air in our present enclosure the fact is that we still have about 1 per cent of post-operative infection and in these circumstances this was very puzzling. It seems impossible that it could come from an airborne route. The fact that the air was clean presented an opportunity for a study of the sterility of the exterior of the surgeon's gown at the end of one hour of performing an operation necessitating considerable muscular effort (total hip replacement). We found (Charnley and Eftekhar 1969 (1)) that even when settle plates in the enclosure remain sterile throughout the operation the surgeon's gown was penetrated by organisms from within, which could be recovered from the surface using the agar-jelly sausage technique. This indicated that the textile used for gowns (balloon cloth) was too thin, and indeed when held to the naked eye inspection revealed visible pores, which were found to reach 50 microns in size. It would seem highly probable that direct contact of certain instruments, and the surgeon's gloves, with the front of the surgeon's gown (as for instance when using a brace-and-bit in his arthroplasty) could explain infection by direct transfer to the wound in the presence of a sterile environment.

It is the porosity of balloon cloth which is responsible for keeping a surgeon tolerably cool in ordinary operations, and the wearing of gowns of thicker and closer weave (ventile) even in ordinary operations can be intolerable for the surgeon. For operations involving considerable muscular exertion heat corresponding to at least 500 watts of electrical energy would have to be eliminated if the gowns were impervious. At the moment we are experimenting with a totally investing thick gown with an internal ventilation system, but it is too early to report on the bacteriological success of its use. They are highly successful from the point of view of the comfort of the wearer.

The future of sterile environments in the operating theatre

My opinion on the future of total asepsis in the operating room is that this will be achieved by a combination of the following criteria:

1. Extraction of body heat, and air carrying infected dust particles from inside the surgeon's gown, by a suitable vacuum system.
2. Wearing of gowns of material impermeable to bacteria, to eliminate direct transfer to the wound (closely allied with the vacuum system in (1)).
3. The use of an enclosure with a pressure gradient above the surrounding theatre to separate the surgical team from ancillary personnel and to permit unrestrained activities of this personnel in the interests of speed and turnover during the operating session.
4. Reduction in the volume of filtered air required by the enclosure as a result of stopping the emission of bacteria at source by (1).
5. Retention of conventional theatre lights inside the enclosure, because laminar flow in this condition is a spurious concept and turbulence is now unimportant due to the elimination of bacterial emission at source by (1).

Results of operating in a sterile environment

I have already said that events are too recent for me to give a final opinion on the relation between post-operative infection and an aseptic environment in the operating room. The complexity of the problem (due to the introduction of a number of additional technical matters, such as improved wound closure) has been described at length in another paper (Charnley and Eftekhar 1969 (2)). I certainly have to agree that sterile air is not the sole answer to the problem of post-operative wound infection, though I believe it has played a very important part in reducing our infection rate from the 5 or 3 per cent level to its present 1 per cent. Nevertheless, I still strongly advocate measures to operate in sterile air for those types of surgery when an extreme asepsis is imperative, and where the organs to be operated on are not viscera already containing pathogenic organisms. There can be no doubt at all that patients are infected from time to time in conventional theatres by carriers of staphylococci in the surgical team (as for instance the recorded instance of an anaesthetist, Payne 1967 (3), and if this is so why should not visitors and other transient personnel be dangerous? The infection rate coming from unpredictable exposure to sources of bacteria from a transient population of visitors, etc., is something which is not subject to statistical analysis. The diagram (Figure 2) indicates the number of wound infections implanted per 100 patients, recorded serially in blocks of 100. These infections did not all present immediately in the post-operative phase, and some manifested themselves as long as 12 months after the operation. (The late appearance of infection after massive implant surgery is included in our infection rate, but if excluded it could reduce recognized infection in hospital to less than $\frac{1}{2}$ per cent.) It will be seen that adjacent blocks of 100 cases could vary in infection rate from zero to 4 or 5 per cent. It is obvious from this that one could not rely on a difference of 1 per cent in infection rate in this type of surgery on less than 300 operations, and a difference of $\frac{1}{2}$ per cent could very well require 600 to 1,000 operations.

Wound closure

The complexity of the problems of tracing the cause of wound infection occurring after operating in clean air is clearly one of great complexity, and it will require another two years for us to test the effectiveness of using impervious gowns in addition to our present clean-air technique. During the next two years it is possible that improvements in the technique of closing the wound, which have obscured our analysis of the effects of clean-air operating during the last two years, will then have clarified themselves.

Conclusions

I think that during the last two and a half years at Wrightington, over some 1500 operations of total hip replacement, we have reached the ultimate in clean-air operating, and this has pointed ways of achieving this at a much lower cost. The clean-air environment has not totally abolished post-operative wound infection, but there is every reason to believe that the residual infection comes from direct permeation of organisms through the textile of the operating gowns, and that it occurs by direct transfer from the gown to instruments or gloves and not via the air.

The architectural details necessary to achieve extremely high degrees of cleanliness in the environment are now clearly defined since the fallacies in the concepts of 'downward displacement' and 'laminar flow' have been recognized.

My experience indicates that a clean-air environment (*a*) can be installed very cheaply, (*b*) does not interfere with the through-put of a busy theatre, (*c*) adds to the comfort of the surgeon, and (*d*) facilitates the post-graduate teaching of surgical techniques, and I think these are excellent grounds for urging its widespread adoption.

BIBLIOGRAPHY

1. Charnley, J. and Eftekhar, N. (1969). Penetration of gown material by organisms from the surgeon's body. **i,** 172.
2. Charnley, J. and Eftekhar, N. (1969). Post-operative infection in total arthroplasty of the hip with special reference to the bacterial content of the air, *Brit. J. Surg.*—accepted for publication February 1969.
3. Payne, R. W. (1967). Severe outbreak of surgical sepsis due to *Staph. aureus* of unusual type and origin, *Brit. med. J.*, **ii,** 17.

Figure 1

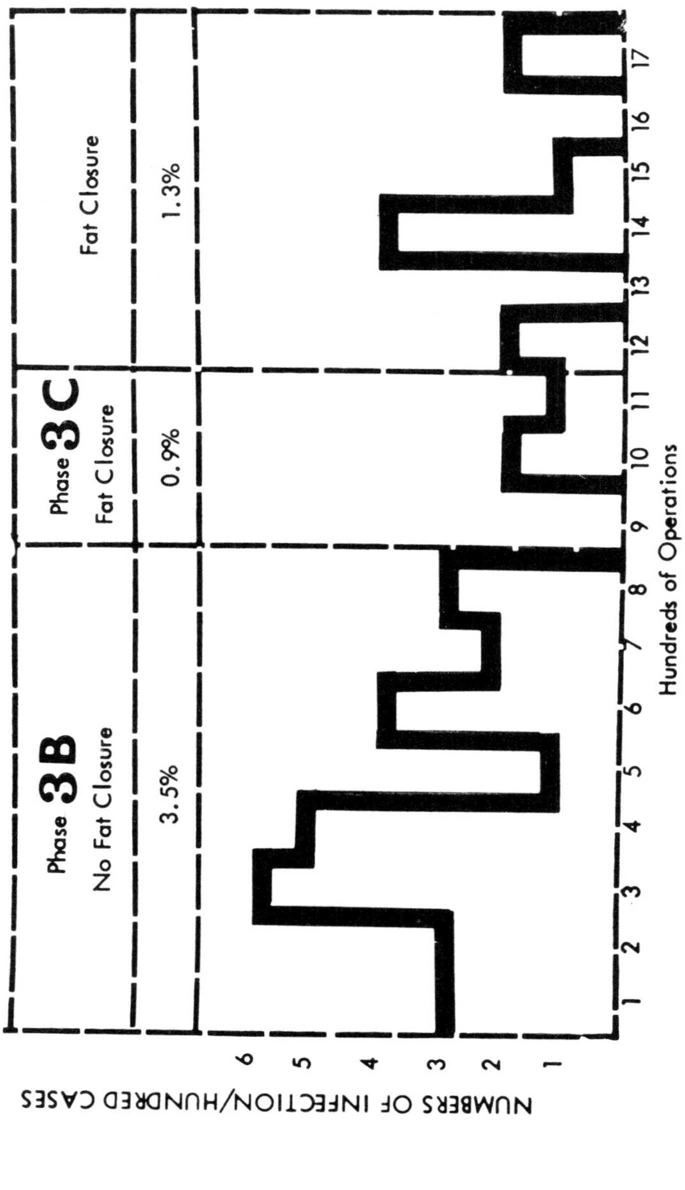
Phase 3B
No Fat Closure
3.5%
Phase 3C
Fat Closure
0.9%
Fat Closure
1.3%
1 2 3 4 5 6 7 8 9 10 11 12 13 14 15 16 17
Hundreds of Operations
1 2 3 4 5 6
NUMBERS OF INFECTION/HUNDRED CASES

Figure 2

Vaccination against anthrax

H. M. Darlow

Microbiological Research Establishment, Porton

Since the first attempts by Toussaint in 1880 a variety of methods have been employed to induce immunity to anthrax, ranging from live spores of avirulent strains to immunogenic materials derived from virulent strains.

Based on the work started by Gladstone thirty years ago, an effective antigen has been developed at M.R.E., which appears to contain the three 'toxin' fractions later isolated by Smith and his colleagues.

This vaccine has been in use in the laboratory for human immunization for 17 years and is now generally available to industries at risk. It has been possible to carry out serological studies on hundreds of recipients during this period, and whilst it has been shown that satisfactory levels of immunity are demonstrable in most cases, there are still some unexplained anomalies worthy of further investigation.

Living attenuated respiratory virus vaccines

D. A. J. Tyrrell

Common Cold Research Unit, Harvard Hospital, Salisbury

There are at the moment no really satisfactory respiratory virus vaccines. Live vaccines present the following possible advantages—they may require smaller amounts of virus to protect, they may be administered without injections, they stimulate the production of local immunity, they may protect also, at least for a short time, by inducing local interference.

The main difficulties are in selecting a virus which is immunogenic and fails to produce symptoms. In the case of influenza viruses it was suggested that simple egg passage would attenuate the virus suitably, but this has not been confirmed. However, it was recently confirmed that mutants of influenza A2 which are resistant to horse serum are quite well attenuated for man. This method is not, however, likely to be generally useful or adequate and there is therefore much interest at the moment in the possibility of using 'cold' or *ts* mutants. Some experimental work in this field will be reviewed. In the case of adenovirus type 4, a strain which would not be attenuated if given into the respiratory tract has produced immunity without symptoms when given orally in enteric-coated capsules.

Current vaccination procedures against influenza

D. Hobson

Department of Bacteriology, University of Liverpool

The epidemic transmission of influenza is too rapid for control by quarantine measures or by chemotherapy. Active immunization in advance of the epidemic wave can control outbreaks efficiently only if certain essential requirements are met.

1. Antibody against the viral haemagglutinin (HA protein) must be induced in order to neutralize virus infectivity.
2. The composition of vaccines may have to be changed at short notice. Influenza A viruses have shown successive antigenic variations over the last 30 years. HA-inhibiting antibody induced against one strain may fail to neutralize subsequent epidemic strains.
3. Even when the vaccine strain, and epidemic strain are in close antigenic relationship, the protection rate against clinical infection in controlled trials may vary from 90 per cent to less than 30 per cent. The frequency and duration of antibody response in a community can be improved by immunological adjuvants, together with purification and concentration of HA antigen.
4. Injected killed vaccines may induce high levels of circulating antibody but fail to stimulate IgA antibody in respiratory secretions, unlike the natural infection or experimental intranasal vaccination with live attenuated viruses.

Adjuvant effects of *B. pertussis* preparations in respiratory immunization of mice with tetanus toxoid

H. C. Bartlema, J. L. F. Gerbrandy and **R. Karelse**

Medical Biological Laboratory of the National Defence Research Organization TNO, Rijswijk ZH, The Netherlands

Inhalation of plain tetanus toxoid in high doses fails to induce an appreciable immune response in mice; however, when inhaled in combination with *B. pertussis* as an adjuvant, much lower doses will confer a marked degree of immunity.

Determination of the 50 per cent protective doses (ED_{50}) of toxoid with and without adjuvant by the respiratory and by the subcutaneous route revealed that the adjuvant activity is much more pronounced in immunization by aerosol treatment than by injection.

A possible explanation for this phenomenon is based on further studies indicating that inhalation of plain toxoid induces partial unresponsiveness to this antigen. This is probably due to the fact that this way of administration implies a much more gradual and prolonged resorption of the antigen than any other way. This resorption might also involve a process of biological filtration, resulting in the removal of that part of the antigen which represents its 'intrinsic adjuvanticity'.

It is considered that, in s.c. vaccination, *B. pertussis* is acting only as a 'conventional' adjuvant whereas, in respiratory immunization, this action is superimposed on its 'extrinsic adjuvanticity', which leads the filtered toxoid to induce immunity rather than partial tolerance.

Aerosol immunization against Rubella using the attenuated vaccine

R. H. Waldman, R. J. Hildebrandt, M. L. Schulkind and F. M. Wigley

Department of Medicine, University of Florida, Gainesville

There are both practical and scientific reasons why the aerosol route of immunization against rubella might be preferable. The practical advantage is ease and comfort of administration. The scientific advantage concerns the finding that a significant percentage of volunteers immunized subcutaneously and subsequently challenged with wild virus were protected from clinical illness and viraemia, but wild virus could be isolated from throat swabs 10–12 days following challenge. This might be explicable on the basis of lack of respiratory secretion antibody to rubella following subcutaneous immunization.

Seventy seronegative persons were immunized either subcutaneously or by aerosol with the Cendehill, rabbit kidney grown, attenuated rubella vaccine. Sera and nasal secretions were obtained for antibody determinations. Volunteers were observed for clinical side-effects of the immunization. Seronegative household contacts were also evaluated serologically and clinically.

The results indicated a greater than 90 per cent seroconversion rate in both the subcutaneous and aerosol groups. The height of the antibody response was the same in both groups. In the aerosol group about two-thirds of the vacinees developed detectable nasal secretion antibody as compared to about one-third of the subcutaneous group. There was no spread of vaccine virus to household contacts in either group, and in neither group were there any vaccine side-reactions other than minimal temperature elevation in a few individuals. There was no case of arthralgia or arthritis in any of the volunteers, despite the fact that this included six adult females in both groups.

The mechanisms of immunity to influenza virus infection

C. H. Stuart-Harris

Department of Medicine, the Royal Hospital, Sheffield

It has been known for 25 years that inactivated influenza virus vaccine will confer a significant degree of immunity against influenza in the field. The mechanisms of this immunity has traditionally been believed to be mediated by antibody formation and the recent experimental work on small numbers of persons confirms that these antibodies appear after subcutaneous inoculation in both serum and nasal secretions. The evidence collected by workers, with viruses such as parainfluenza 1, has shown that both serum and nasal antibodies are required to confer resistance to direct challenge by intranasal virus. The experiments made in Chicago with parainfluenza type 2 virus by Tremonti *et al.* also suggest a synergistic effect of nasal and serum antibody. Our own work with influenza B virus appears to give somewhat similar results. At the same time our work has emphasized the great difficulty in obtaining consistent results in volunteers, perhaps because of changes in the immune status of the population resulting from natural exposure to virus. Thus the ready demonstration of nasal antibody before immunization in the volunteers in 1968 compared with its absence from others in 1967 is strongly suggestive of recent exposure to influenza B virus. The results of attempted infection also tally with exposure to influenza B virus in the preceding months of 1968. Further work with volunteers is clearly required using different persons and different virus strains in order to establish the role of nasal and serum antibodies in relation to natural and experimentally induced immunity against influenza viruses.

Antigenic properties of influenza virus proteins

V. Pavilanis *et al.*

Institute of Microbiology and Hygiene of the University of Montreal

The low efficacy of current influenza virus vaccines, and the observation of local and general reactions after inoculation of this vaccine, indicate a need for a vaccine of high potency that can be given in large doses to persons of all ages without causing side effects. For this purpose we studied the antigenicity of purified influenza virus proteins.

A/PR8/34 influenza virus strain was concentrated by adsorption onto chick red blood cells and elution with RDE. Then the virus was treated following Davenport's method with Tween-ether and lanthanum acetate. By this procedure, it was possible to obtain only virus proteins. Examining the suspension in the electron microscope, we did not observe any virus particles.

The purified influenza virus proteins inoculated into guinea pigs produced a good haemagglutination inhibiting (HI) antibody response.

The purified influenza virus proteins, diluted to contain 22 CCA, were inoculated into 13 adults. No general or local reactions were observed in the vaccinated persons. The geometric mean of the titre of HI antibodies before vaccination was 1/112. Two weeks after, the geometric mean of the antibody titre was 1/1066. All volunteers responded with a more than fourfold increase in antibodies. Four weeks after single inoculation of the vaccine, the geometric mean of the titre reached 1/1474. There was no change in antibody titre against A2/Hong Kong/68 strain.

Seventy-seven per cent of those vaccinated showed an increase of CF antibodies against the viral antigen of A/PR8/34. Average titres were 1/32 two weeks after and 1/64 four weeks after vaccination. No increase of CF antibodies against the viral antigens of A2/Australia/57 or A2/Hong Kong/68 was observed. Only 30 per cent showed a twofold increase of CF antibodies against the soluble influenza A antigen.

Vaccines and methods of vaccination

Résumé: presented by **F. T. Perkins**

The four papers concerning immunization against influenza summarized our knowledge about:

(i) the correlation between immunity and levels of antibody,
(ii) the part played by local and circulating antibody,
(iii) the immunogenicity of fractions of the influenza virus, and
(iv) the efficacy of live attenuated vaccines.

Dr. Hobson reminded us that an attack of influenza virus gives rise to HI antibody, the titre of which appeared to correlate well with protection against further episodes. It was pointed out, however, that this may be no more than an index of immunity and that other antibodies may also play a significant part. Many of the relevant studies were done some years ago and more recent findings suggest that locally produced IgA antibody or antineurominidase may be as important as the IgG produced in the circulation as a result of infection.

The appearance of different serotypes of the virus continues to hamper rapid progress in the field of influenza immunization. The philosophy concerning the composition of the vaccine continues to be a subject for interminable debate. It was clear that vaccines devoid of the virus of the epidemic type would not protect against the disease, whereas monovalent vaccines of the epidemic type were about 60–70 per cent effective.

During the general discussion it became clear that the current research into splitting the virus with desoxycholate or extracting the haemagglutinin with a detergent and solvent were increasing our knowledge of the antigen giving good antibody responses. There was need for a great deal more work, however, since all strains may not behave in a uniform manner when treated with chemicals. Furthermore, it was important to ensure that the antigen responsible for protection was being extracted. There was a note of caution concerning the procedures of adsorption and elution used for the purposes of purification of the virus harvest. When chick red cells were used, for example, appreciable amounts of chick red cell antigens appeared in the elluate which would be an undesirable component of vaccines.

The evidence that the haemagglutinin extracted from whole virus gives a good antibody response in laboratory animals and man is most encouraging. The reports that such preparations give rise to fewer reactions in man, however, are being questioned, because in two small trials there appeared to be no difference in the local reactions to the whole virus or extract antigen vaccines. This needs further investigation.

Professor Stuart-Harris reported studies on the appearance of antibody in nasal secretions after vaccination which are now being followed with interest. Already the suggestion has been made that some vaccines or vaccination procedures, though producing circulating antibody, fail to produce antibody in the nasal secretions. Both may have a synergistic effect and both may be necessary for protection against reinfection. As with previous work using parenteral inoculation of killed whole virus, these studies are further confused by the previous immune status of the subjects taking part in the trials, but much effort should go into furthering these studies, using larger groups of volunteers.

The relative efficacy of live attenuated vaccine compared with killed vaccine was also considered. Dr. Tyrrell revealed that strains made resistant to horse serum inhibitor became attenuated for man, but problems inherent to the selection of an attenuated strain need further work. The temperature of incubation and the number of passages for attentuation were of immediate importance, as were the dose of virus required and method of application for successful immunity. When protection trials were carried out both live vaccine given intranasally and killed vaccine given parenterally produced antibody, but on intranasal challenge there was better protection in those subjects given the live vaccine.

The general discussion revealed that research concerning the production of purified and killed whole virus vaccines as well as antigen extract vaccines should continue in parallel with the development of live attenuated vaccines. In the production of killed vaccines special attention should be given to the particular antigen extracted, making sure that this was indeed the 'immunizing' antigen. An efficient adjuvant to give better and longer lasting protection with the use of less antigen would be most helpful in this regard. In the production of a live attentuated vaccine the technique of attentuation and marker tests to assess attenuation were of outstanding importance. There was some concern that the time taken to prepare and safety test a live vaccine would be too long to be of any help in the face of a pending epidemic.

More data on the efficacy of vaccines given by the intranasal route were presented by *Dr. Waldman,* who has studied rubella vaccine. It has been shown that although subjects given live attenuated rubella vaccine by the parenteral route are protected against viraemia and the appearance of clinical symptoms, no less than one-third of them shed virus from the pharyngeal area when challenged with live virus. A live attenuated vaccine made from the Cendehill strain in rabbit kidney cells was given to children either by aerosol spray or by the subcutaneous route and the HI antibody responses were measured both in sera and nasal secretions. Both methods of vaccination gave antibody rises in a similar percentage of subjects ($>$ 90 per cent), but the rises after aerosol administration were greater. The percentage of children having antibody in nasal secretions, however, was much higher (70 per cent) in the aerosol-vaccinated children than in those given vaccine parenterally (30 per cent). There were no reactions to the administration of this vaccine given as an aerosol.

Still further data concerning the nasopharyngeal region as a route of vaccination were presented by *Dr. Bartlema,* who reported a study of tetanus toxoid in mice. It was shown that tetanus toxoid given as an aerosol did not stimulate the production of antibody. Mixing the tetanus toxoid with *Bordetella pertussis* organisms, however, gave high titres of circulating tetanus antitoxin. The mechanism of this adjuvant effect was unknown and more work was needed to determine whether this phenomenon was unique to *B. pertussis* or whether some component such as the toxin of this organism would also be effective. The observation, however, is worthy of further study with other antigens.

Finally, it was encouraging to learn from *Dr. Darlow* of the successes achieved at the Microbiological Research Establishment, Porton, in immunization against anthrax. The extraction of the immunizing antigen from an avirulent strain grown in a specially designed medium has yielded the non-toxic, non-reactive Belton/Strange antigen that gives rise to antibody responses and protection in man after three injections. The vaccine has been given inadvertently to subjects who failed to report a history of anthrax and there were no untoward sequelae. A safe and effective vaccine is therefore available, and since there are no reactions to repeated injections, booster doses are recommended for those people at special risk to the disease. Although the development of the vaccine is a major step forward in immunity against anthrax there is a need for the development of a potency test in a laboratory animal.

Relationship between environmental temperature and the survival of airborne bacteria

Richard Ehrlich *et al.*

IIT Research Institute, Chicago, Illinois

Studies reported in this paper were concerned with the effects of environmental temperatures ranging from −40° to 120°F on the behaviour of airborne *Serratia marcescens*, *Escherichia coli*, and *Bacillus subtilis* var. *niger*.

Aerosol recoveries and biological death rates of *B. subtilis* spores were not significantly affected by the environmental temperature and remained approximately constant within the temperature range studied.

The survival of airborne *S. marcescens* and *E. coli* was closely related to the environmental temperature. An increase in temperature from 0° to 120°F resulted in a significant increase in the biological death rate. Within this temperature range the relationship between the biological death rate and the environmental temperature appeared to be linear. An increase in temperature from 75° to 120°F resulted in significantly reduced aerosol recoveries of the two vegetative micro-organisms 4 minutes after dissemination. However, the differences in aerosol recovery observed at temperatures ranging from 0° to 75°F were not significant. At −40° and −20°F the aerosol recoveries of all three agents were consistently lower than at 0° to 75°F, although the death rates were not significantly altered.

On the mechanism of the decay of poliomyelitis virus and encephalomyocarditis virus in aerosols

J. C. de Jong

Laboratorium voor Microbiologie, Utrecht

Poliovirus and EMC virus were sprayed into a static air chamber at various relative humidities and were recovered with impingers. The samples were titrated for infectious virus, for hemagglutinating activity (EMC virus) and, after extraction with phenol, for infectious RNA (poliovirus and EMC virus).

EMC virus was grown in L cells in Eagle's medium without serum and suspended after ultracentrifugation in Hank's balanced salt solution without $NaHCO_3$ at pH 7·2. One ml. containing 10^{10} p.f.u. was directly sprayed into a static air chamber at 20°C. The aerosols were sampled with the lower stage of the May impinger, containing phosphate buffered saline. The samples were titrated for infectious virus, for hemagglutinating activity and, after extraction with phenol, for infectious RNA (efficiency with normal virus: 10^{-2} to 10^{-4}).

Results with EMC virus. The virus rapidly lost infectivity (up to three logs) during spraying at relative humidities below 60 per cent. After spraying the virus was rather stable, except between 50 and 65 per cent RH. At 40 per cent RH the hemagglutination titre closely followed the infectivity, whereas the RNA within the virus particle remained infectious for at least one hour. Aerosolized free infectious RNA also was stable. At 70–90 per cent RH neither the viral components nor the infectivity was impaired; again sprayed infectious RNA was stable.

From the close relationship between hemagglutination and adsorption to susceptible cells with EMC virus one can infer that, at least at 40 per cent RH, the cause of inactivation of EMC virus in aerosols is *deterioration of the protein component.* Whether the RNA is freed thereafter from the protein coat is not known.

In earlier experiments poliovirus was sprayed in Hank's balanced salt solution with calf serum 2·5 per cent, lactalbumin hydrolysate 0·5 per cent and peptone 0·5 per cent. Aerosols were sampled with raised all-glass impingers containing Dulbecco's PBS with peptone 1 per cent and antifoam. The impinger fluids were concentrated by ultracentrifugation and titrated for virus and, after extraction with phenol, for infectious RNA (efficiency with non-sprayed virus: 10^{-3} to 10^{-7}).

Results with poliovirus. The dependence of inactivation on RH was similar to that with EMC virus. Oxygen is not an important factor, as was demonstrated by experiments in nitrogen. The RNA within the virus particles was found to decay parallel with the whole virus at three different RHs. The cause of the discrepancy between this result and that with EMC virus is not known, but technical differences may be responsible (perhaps involving the presence of RN ase in the experiments with poliovirus).

The Open-air Factor

H. A. Druett

Microbiological Research Establishment, Porton

The use of the microthread method to expose organisms to the open air led to the observation that the survival of bacteria in the night air of Porton in no way resembled that predicted for these ambient conditions by laboratory methods such as the rotating drum. The Open-air Factor varied considerably from night to night, being greatest under meteorological conditions conducive to higher levels of pollution. In considering this problem experiments were made with a range of pollutant gases such as sulphur dioxide, oxides of nitrogen, formaldehyde, butylaldehyde, carbon monoxide, and with ozone. In addition, a few experiments were made with primary pollutant sources. Various physical parameters were also considered. While these experiments were designed to screen the respective pollutants as candidates for the cause of the open-air factor, the data may be of interest to other investigators in a different context.

There appears to be a strong prima facie case for relating Open-air Factor to pollution from motor vehicles, but contributions from other sources should not be ruled out.

Airborne transmission potential of murine leukemia infection

Edgar W. Larson *et al.*
Fort Detrick, Frederick, Maryland

In mouse colonies maintained for experimental purposes, leukemia infections appear to occur spontaneously with apparent transmission of the disease between cage mates and from cage to cage. These observations suggest a potential for airborne transmission. The objective of this work was to obtain preliminary information on the aerosol stability of Rauscher murine leukemia virus. Aerosols of the virus (No. HL310) were generated in a static cloud chamber conditioned at 75°F, 40 per cent relative humidity, approximating to the conditions found in animal colonies. Aerosol samples were collected at cloud ages of 9 and 35 minutes post-dissemination. Virus assays were performed by intraperitoneal injection in mice with increased spleen weight indicating response. Not more than 20 per cent of the virus was inactivated by sampling processes *per se*. A high proportion of the virus was inactivated, however (over 99 per cent), in the process of aerosolization and equilibration with the environment. Thereafter, the virus aerosols were relatively stable, decaying at approximately 1·0 per cent per minute. These studies demonstrated the feasibility of applying conventional aerobiological techniques in the study of mouse leukemia virus. Insofar as is known, the work constitutes the first investigations directed to the quantitation of aerosol stability with these viruses. Important extensions of the work will include the direct exposure of animals to aerosols and investigation of virus suspension and environmental variables on aerobiological properties.

Microbial survival in air

Résumé: presented by E. K. Wolfe

The first paper dealt with bacterial survival over a wide range of environmental conditions. Bacterial survival has probably received more study than any other aspect of aerobiology; techniques are well established and widely known and used, but only now are comprehensive data becoming available through such experimental work as was described by *Dr. Ehrlich.* Part of the difficulty met in conducting such comprehensive studies of this kind is the expense and the technical problems in determining the best method for even such a simple procedure as sampling. However, it is gratifying to state that the only real problem at the present stage is that of the time required to carry out these investigations.

The next two papers related to the survival of viruses. Here the path is much less well charted than for bacterial survival. The technical problems are more severe, assay procedures are expensive, often imprecise, and frequently need much more time than is required for bacteria. One hesitates to place too much confidence upon the implication of results of a single experiment even though it consisted of several replications and required several weeks to complete. *Dr. de Jong's* studies of the mechanism of decay of poliomyelitis and EMC viruses indicate the magnitude of the problem since the RNA inactivation in relation to whole virus inactivation was completely different in the two viruses. *Mr. Larson*'s studies of the survival of the Rauscher mouse leukaemia virus will certainly be no source of comfort to Dr. Hellman, who is concerned with the biohazards of aerosolized oncogenic agents. The paper is indicative of the growing interest in the possibility of airborne transmission of cancer-producing agents and shows the need for a considerably increased effort in this field.

Only in recent years has much attention been given by aerobiologists to the biocidal activity of minor or trace components of the atmosphere either natural or man-made. *Mr. Druett* very ably collated the data on bactericidal activity of these components and showed that most of the well-known pollutants were of little or no significance. These bactericidal components are effective in such minute quantities that the presence of a living organism such as a bacterium is required to demonstrate their existence since to date all analytical procedures have failed to detect them.

The capability for rapidly producing useful data in these investigations was made possible by the May/Druett microthread technique, but paradoxically has raised a major problem. MRE* unquestionably has the most experience and greatest expertise in working with this technique. The U.S. and Canada are now also using the technique and they not only fail to confirm the M.R.E. results but they also fail to agree with each other. In Holland investigations using this technique have started and similar problems seem to have been encountered. In each country the effort has been made to duplicate the M.R.E. equipment and procedures, but it is clear that sufficient differences exist to markedly affect the results. Meetings such as this, giving opportunity for direct discussion, will play an important role in resolving the problems of standardization. That these can be resolved is evident in the fact that in aerosol studies in closed containers the various countries can obtain remarkably similar results. As indicated above these meetings serve a valuable purpose, not only in bringing aerobiologists into contact with investigators from other disciplines with common areas of interest but also in bringing aerobiologists into contact with aerobiologists having common problems. At the first international symposium in 1963 the all-glass impinger sampler was adopted as the standard for all aerosol investigations. The papers presented at the 1966 meeting confirmed the value of standardized equipment. It is to be hoped that the 1972 conference will show that satisfactory standardization has been achieved with the microthread technique.

* The Microbiological Research Establishment, Porton.

Rapid diagnosis

N. R. Grist

Regional Virus Laboratory, Ruchill Hospital, Glasgow

Standard diagnostic laboratory procedures for viral and bacterial infections take hours, usually days, occasionally weeks or longer. 'Rapid' methods are those which reduce hours to minutes and weeks to days or hours. Standard methods of rapid laboratory diagnosis are as follows:

1. Visualize the infecting organism, provided it is available in adequate quantity in a specimen and that its appearances are sufficiently characteristic. Staining the organisms is helpful for optical microscopy. Viruses may be seen by electron microscopy of phosphotungstate-stained specimens.
2. Detect specific antigen of the organism, provided it is available in adequate quantity in a specimen, and that satisfactory specific antiserum is available. Methods include staining with fluorescent antibody, agar gel precipitin tests, and complement fixation.
3. Cultivate and identify the organism, provided it is available in viable form in a specimen. This requires supplies of appropriate culture medium—usually nonliving fluid or gelled media for most bacteria. Viruses and some fastidious microbes require tissue or organ cultures or animals or embryonated eggs. Growth and identification take significant time—usually days, perhaps weeks or occasionally months. Rapidity may be gained by recognition of early growths or antigens by such methods as immunofluorescence.
4. Measure the patient's antibody response. Without waiting for a week to demonstrate rising titres in serial sera, rapid diagnosis may sometimes be achieved by tests for particular antibodies (e.g. IgM) characteristic of the early but not late stages of immune response.

Rapid methods develop in response to demand, as exemplified by the public health urgency of smallpox and the recent therapeutic urgency of herpes encephalitis in which the use of idoxuridine may be important. Both of these exemplify the incorporation of new rapid methods into an established battery of standard methods which must be retained for their greater sensitivity and reliability. Gas chromatography is one example of a possible new approach to rapid and early (possibly pre-symptomatic diagnosis not based on the traditional methodology, and developments along new lines such as this may be important.

Microcolony techniques in rapid diagnosis

P. Chadwick

Department of Microbiology, Queen's University, Kingston, Ontario

---

Routine bacteriological diagnosis, including full characterization of a pathogenic organism, involves one or more periods of overnight incubation. Use of microcolonies as material for investigation promises to accelerate bacteriological diagnosis, as observations can be made within a few hours of initiating cultures.

Impressions of microcolonies on cover-slips may be treated with specific fluorescent antibody. The flourescent microcolonies in these preparations can be seen by low-power microscopy, and this method promises to be of high sensitivity. Alternatively, microcolonies of some bacteria may be treated with specific bacteriophage and their subsequent disintegration observed microscopically.

In very few instances will microcolony structure reveal bacterial identity, but useful information may be derived from growth of microcolonies of certain size and features on specific selective media.

Rapid diagnosis can include the antibiotic sensitivity spectrum of an organism, if a method is used whereby resistance or sensitivity to an antibiotic is measured in terms of the presence or absence of microcolonies on antibiotic agar after four hours' incubation.

The various microcolony procedures may be used in appropriate combinations. The scope of such methods in diagnosis of airborne disease remains to be defined, but the methods are simple and possess the virtue that micro-organisms are examined by direct observation.

The application of immunofluorescent antibody techniques to the rapid diagnosis of virus infection

P. S. Gardner

Department of Virology, Royal Victoria Infirmary, Newcastle upon Tyne

The ultimate purpose of rapid diagnostic procedures is the control and management of virus disease. The immunofluorescent technique must be comparable in sensitivity to current diagnostic procedures. This means that specific reagents and adequate controls are always employed.

Patients' specimens must contain intact cells so that specific fluorescence will be intracellular. Tests may be made on such specimens or on inoculated tissue cultures.

The indirect fluorescent antibody technique has been used throughout and this communication describes those viruses which have been investigated in this Department by this method.

Respiratory syncytial virus, the most important pathogen of children, is responsible for some 60 per cent of respiratory diseases and is a common cause of death in those under one. Results have shown that of 103 nasopharyngeal secretions from which RSV was subsequently isolated, 99 were diagnosed on the day of admission by fluorescence. This test could be applied to infected tissue culture for confirmation of type of virus present.

Parainfluenza viruses. The same principles have been applied but the problem is complex, with antigenic crossing amongst these viruses and with mumps and SV5.

Influenza A. Liu and Hers originally established the reliability of the fluorescent technique for rapid diagnosis of influenza; this we confirm.

The indirect fluorescent antibody technique has proved to be a useful tool for investigating the lungs of children who die of infections involving respiratory viruses.

Herpes virus hominis. The method has been extended to the investigation of corneal ulceration and skin lesions suspected of being herpetic. The rapid diagnosis of this group of viruses is important because of their sensitivity to idoxuridine.

Experience with a mobile laboratory unit: a preliminary report

J. F. Ph. Hers

Department of Clinical Virology, University Hospital, Leiden, Netherlands

The Department of Clinical Respiratory Virology has been equipped with a mobile laboratory unit since September 1968 and has been operating in an area covered by a radius of 150 km. around Leiden. This Unit contains incubators for tissue cultures and low-temperature cabinets and is completely self-contained. A radio-telephone set, covering the whole country, allows personal consultations with the Department, hospitals, general practitioners, doctors and patients. The unit was developed

(1) to facilitate the isolation of respiratory viruses from patients outside the hospital, avoiding negative results caused by transport difficulties of the material (sputum, swabs, throat washing, autopsy material),
(2) to accelerate rapid information of outbreaks of respiratory illness in the population and segregated communities (army camps, schools, creches or day nurseries), and
(3) to perform virological and bacteriological studies at the subclinical level of respiratory minor illnesses.

Rapid information on the outbreak of Hong Kong influenza (virus A2) was recorded in the population, while an outbreak of infection caused by respiratory syncytial virus was diagnosed twice in two segregated communities of 3–6-year-old children.

A follow-up study on subclinical respiratory minor illnesses was started in a children's creche in The Hague. Of the 250 children 2–5 years old, 50 showed minor and trivial symptoms. In 14 of these 50 children a virus diagnosis could be made, 5 with influenza virus A2, 5 with respiratory syncytial virus, 1 with adenovirus and 1 with *Mycoplasma pnumoniae.* In two additional cases serologic evidence of concurrent infection with adenovirus was found, showing a more than fourfold rise of specific antibody. Mumps virus was isolated in one of the adult nursing staff. Infected monolayers from the mobile unit were screened for specific fluorescence 24 and 40 hours after inoculation by the indirect immunofluorescent test; autopsy material (impression smears and ergostatic sections) and preparations of sputum and/or nasal exudate when available were also stained for specific viral antigen. A group-specific diagnosis proved by virus isolation could be made in nearly all cases within 48 hours. The results obtained so far show that because of the effectiveness of collection of material by a mobile unit a higher isolation rate of respiratory viruses can be obtained and rapid sensitive diagnostic methods, e.g. immunofluorescence, also become more effective. Some of these specimens, stored in transport medium, were also sent by mail or transported with the mobile unit to the laboratory and then inoculated. Bacteriological cultures were included in all cases.

Rapid diagnosis in biological defence

J. F. Metzger
U.S. Army Medical Unit, Fort Detrick, Frederick, Maryland

The specimens to be utilized for rapid diagnosis in a biological defence concept must be obtained from the environment or from a 'healthy' military population. Unlike the specimens received in a clinical laboratory practice, the history of illness will not be available.

Since biological agents may be disseminated by contamination of the environment, air and water samples are of prime importance. An ideal system would be one that automatically monitors the environment on a continuous basis.

Research activities of U.S. Army Medical Research Institute of Infectious Diseases have been centred on the immunological, biophysical and biochemical changes which occur during the incubation period of a variety of infectious processes.

Immunofluorescent identification methods have been studied extensively. Utilization of this technique is limited in actual practice if no clues are available as to which group of organisms may be involved. Cross-reactions between related organisms often occur. Utilization of the indirect test for detection of antibody requires again historical information as to which antigens should be utilized in the tests.

Infectious agents can be separated from the sample by biophysical methods and demonstrated by the use of the electron microscope. Immunologic procedures such as ferritin-labelled antibody can be utilized as an identification adjunct to this procedure.

Biochemical changes occurring in serum or plasma during the incubation period of a variety of infectious diseases have been investigated actively by the Physical Sciences Division. These include changes in trace metals, amino acids, tissue enzymes, glycoproteins, and hormones.

In summary, rapid diagnosis must eventually depend on procedures which do not require specific immunological reagents but rather an analysis of biochemical and biophysical parameters.

Résumé: presented by **J. C. N. Westwood**

The relevance of rapid diagnosis to aerobiology may appear to be marginal, but its importance as a subject for study cannot be doubted.

Critical decisions both in relation to individual patients and to public health preventive measures depend upon accurate diagnosis, and the faster this can be achieved, the more valuable it is. Emphasis must not, however, be exclusively on speed, and in his introductory paper, *Dr. Grist* stressed the importance of both *speed* and *certainty* as equally important aspects of the problem. Dr. Grist outlined the present position of diagnostic virology and gave a balanced picture of the several methods which are available. Rapid diagnosis does not depend solely on virus isolation, but detection of viral antigen by immunological techniques, of virus particles by electron microscopy and the intelligent use of serology all have their place. Inevitably the fluorescent antibody technique with its sensitivity, ability to detect early viral antigen formation and above all its specificity influences all aspects of this work.

Dr. Chadwick continued the story in the field of bacteriology and outlined the technique of microcolony examination three to four hours after the seeding of cultures. While the technique is of limited value in the examination of primary cultures, it is extremely valuable in the rapid determination of antibiotic sensitivities and, with the use of fluorescent antibody staining or of bacteriophage challenge, for rapid species identification. While antibiograms of some fast-growing organisms may be determined in four to five hours by standard macro-methods, this is not so for slower-growing bacteria and it is for these that the method is of particular value. In Dr. Chadwick's laboratory, the microcolony technique is routinely used for this purpose and the sensitivites of all but a few isolates are reported on the day of first isolation—a service greatly appreciated by clinicians. A particularly useful application of the method derives from its ability, in combination with fluorescent antibody staining, to detect very small numbers of a specific type of organism in the presence of a great excess of other bacteria. Enteropathic *E. coli*, for example, may be detected in the faeces of carriers when present in a ratio of only $1 = 10^5$ in relation to other faecal organisms.

In the field of virology, once again *Dr. Gardner* recounted his remarkable success in the application of the fluorescent antibody technique to the diagnosis of respiratory syncytial virus infection in children. Direct examination of aspirated naso-pharyngeal secretions specifically stained by the indirect method reveals virus antigen in characteristic distribution within desquamated bronchiolar cells. The examination may be completed in a matter of hours and has enabled Dr. Gardner to establish a routine yielding accurate diagnosis within 24 hours and virtually complete correlation with the much slower virus-isolation methods. Since secondary bacterial complications are rarely encountered in RSV infection, a fast positive diagnosis is of direct value in determining treatment. Extension of this work to cover influenza and herpes simplex infections is now under way.

These three papers gave rise to brisk discussion on the values and uncertainties of the fluorescent antibody technique. The fact that the decisions are totally subjective and depend on comparing brightness with a memory of brightness is the greatest weakness of the method, but this may be remedied by recently introduced instrumentation for measuring the fluorescence of individual cells. The great value of the method lies in its specific ability to detect early antigen formation and in its sensitivity. The latter gives rise to a practical and philosophical problem in relation to 'false' positive or 'false' negative results. How does one assess a new technique or interpret individual discrepancies when the technique proves to be more sensitive than that which is being used as a 'control' of its reliability? Caution is clearly required and the discussion emphasized that definitive and critical decisions should not normally be based on the results of only a single technical procedure.

The participants in this seminar were all practical microbiologists and fully aware of the fact that there is more to rapid diagnosis than fast laboratory technique. If a specimen requires 24 hours to arrive and the report a further 24 hours to reach the physician, the value of a three-hour laboratory diagnosis is largely negated. *Dr. Hers,* exasperated by the increasing slowness of postal virology and by the necessity for processing 'the wrong specimen taken from the wrong patient at the wrong time', persuaded the Netherlands Government to finance the equipping of a small van as a mobile laboratory. In his paper, Dr. Hers described the equipment of this laboratory, which is designed for the collection, inoculation and storage of specimens, and its radio-controlled operation with a crew of three. The unit does not cover the centre of Leiden itself, since adequate services are available in that area, but serves the outlying areas within a 50 kilometre radius. When called upon for diagnostic assistance in the face of an outbreak, the physician in charge of the crew obtains the names of a number of patients with the syndrome in question and himself selects the cases to be tested and collects the appropriate specimens. These are immediately processed in the unit, inoculated to appropriate cell cultures and placed within an incubator designed for stability and freedom from vibration. Dr. Hers emphasized that this service is primarily a research project to judge the efficacy of such a unit in providing fast accurate diagnosis to physicians and public health services. The project is still in its early stages but has already shown a gratifying increase in the proportion of isolates obtained.

In discussion, it was felt that no judgement could yet be made but the project was an excellent one and deserved success.

The concluding paper by *Colonel Metzger* opened up entirely new ground for consideration. Switching from preoccupation with the infecting agent to the disease process itself, Col. Metzger presented a series of observations on biochemical changes in the body fluids occurring during the first few days after infection in a number of diseases. The substances studied included trace metals, such as copper, iron and zinc, glyco-proteins and amino-acids and were selected largely on the basis of the possibilities of automation of their estimation techniques. This work, too, is still in its early stages but already suggests a very real diagnostic potential. It is likely that most of the changes observed are not specific to individual diseases, but they may be specific to diseases caused by particular groups of agents. For instance, it appears that early serum glyco-protein changes may be used to distinguish between viral and bacterial meningitis. Dr. Metzger's paper raised thought-provoking possibilities in relation to automated computerized diagnosis.

The final discussion which brought the seminar to a close generated two main conclusions about which there was general agreement. First, it was clear from the papers themselves and from the attitudes of discussants that the problems of rapid diagnosis fall into three distinct though overlapping categories:

(i) Research and special application.
(ii) Public health.
(iii) General patient diagnosis.

In the first category, the field of operation is carefully selected and defined and the results suitably controlled. Such projects may be expected to yield a high proportion of positive results and Dr. Gardner's work illustrates this point. Dr. Hers' project by contrast falls clearly into the second category despite being still in the research stage of development. The laboratory physician selects from a number of patients notified to him those most likely to yield a positive diagnosis. Implicit in this system is the fact that the patient is not viewed as an individual problem but as a representative of the ailing public, and the diagnosis sought is that of a disease prevalence. Here again, a high rate of success may be expected from the type of approach described by Dr. Hers and the service could be of immense value.

The final category is that of the service microbiology of the hospital and general practice diagnostic laboratory. Both Dr. Grist's and Dr. Chadwick's papers were related to this field. Here the diagnostic problem concerns individual patients indirectly related to individual treatment. Specimens are taken at the whim of the clinician, with the laboratory exercising no control over their selection or timing, and are transported by whatever means are available and rarely under optimal conditions. This is the area of true 'knock-about' diagnostic microbiology and the uncertainties affect bacteriology and virology in different ways. The yield of useful positive results is usually low and is reduced in direct proportion to the unsuitability of specimens submitted. For instance, virological specimens from children's wards are usually taken too late. Since the earliest possible specimens are required, it may be arranged that throat washings and stools be routinely submitted on admission. The isolation rate from virus infections may be increased in this way, but the overall rate will be greatly reduced by the unavoidable inclusion of totally irrelevant specimens from cases admitted for non-virological conditions. It is this category of work which brings the diagnostic microbiologist to the second main conclusion of the seminar—that rapid diagnosis is as much an administrative as a laboratory problem. To make an effective contribution to increased speed, a technique must normally reduce an 18–24-hour procedure to less than six hours, since otherwise it offers no advantage over an overnight procedure. Most laboratories are fully manned for only 40 out of the 168 hours of each week with at best skeleton staffs in the evenings and at weekends. Since most biological processes work to their own intrinsic time-scale over which we have little control, marked improvement in speed of diagnosis cannot be expected until more adequate staffing of the laboratory services can be obtained.

Differential susceptibility to respiratory virus infection by blood group status

C. W. Potter

Department of Medical Microbiology, University of Sheffield

During the past ten years, a number of reports have appeared of investigations into the association between blood group status and susceptibility to certain virus infections. Infections by Coxsackie A21 virus (McDonald and Zuckerman 1962; Tyrrell *et al.* 1968) and adenovirus types 12, 5, and 6 (Shedden and Potter 1964) have been reported to occur with the expected frequency in persons of blood group A and O. However, acute hepatitis and infections by adenoviruses have been reported to occur in a significantly greater proportion of servicemen of blood group A, compared with those of blood group O (McDonald and Zuckerman 1962; Zuckerman and McDonald 1963), while infections by A2 influenza virus were noted in significantly more persons of blood group O than in persons of blood group A (McDonald and Zuckerman 1962). Other studies, particularly of infections by influenza viruses, have not confirmed the above findings. In the present report, we have examined the suggestion of some authors that the variable results may be due to differences in the immune status of the groups studied.

Infections by influenza viruses

Evidence that susceptibility of man to infections by influenza viruses was related to blood group status was first put forward by McDonald and Zuckerman (1962). In a study of approximately 2000 R.A.F. men admitted to sick quarters with acute respiratory infections during the period 1956–1961, the authors reported that infection by A2 influenza virus occurred more frequently in persons of blood group O than in persons of blood group A ($P = < 0{\cdot}0001$). However, a later study of 221 serologically proven cases of A2 influenza showed no evidence of a higher incidence of infection among blood group O persons compared with control groups (Fagerhol *et al.* 1964). Tyrrell *et al.* (1968) reported that infections of volunteers given attenuated influenza vaccines occurred more frequently in blood group O persons than in persons of blood group A, and this was particularly noticeable in studies with attenuated B/Eng./13/65 virus vaccine.

If the exclusive findings of McDonald and Zuckerman (1962) can be extended to the general population, more persons of blood group O should possess serum antibody to A2 influenza virus, as a result of past infection, than persons of blood group A. To examine this possibility, tests for haemagglutination inhibition (HI) antibody to influenza virus A2/Singapore/1/57 were carried out using 515 serum specimens collected in the period 1961–1963 (Potter and Schild 1967). The results are given in Figure 1. The incidence of HI antibody in serum specimens from persons of blood group O was greater than that from persons of blood group A in each of the seven age groups studied. These differences indicate that past infection by influenza A2 virus had occurred with greater frequency in persons of blood group A ($P = <0{\cdot}01$). Furthermore, the studies of McDonald and Zuckerman (1962) were based on patients admitted to sick quarters and represent an analysis of selected cases. The results shown in Figure 1 indicate that the higher incidence of infection by A2 influenza virus in persons of blood group O was not restricted to cases of severe clinical infection.

The analysis of serum specimens for HI antibody to influenza virus A2/Singapore/1/57 by blood group status was repeated using 369 sera collected in 1966. The tests were extended to include HI antibody to influenza viruses A/Swine, A/PR8 and A1/FM1. The results are shown in Figure 2. The greatest incidence of HI antibody to influenza virus A2/Singapore/1/57 was found in persons aged 16–25 years, to A1/FM1 virus in persons aged 21–25 years, to PR8 virus in persons aged 26–30 years and to A/Swine in persons aged 51–60 years. These results are similar to an independent study reported from the same community. (Schild and Stuart-Harris 1965). Analysis of the results by blood group status, and in seven age groups, indicated that in no instance, individually or collectively, was there evidence that past infections had occurred with a greater frequency in persons of blood group O compared with persons of blood group A. This was true for all four viruses used.

Infection by adenoviruses

McDonald and Zuckerman (1962) reported that, during the period 1956–1961, infections by adenoviruses had occurred in significantly more servicemen of blood group A compared with those of blood group O ($P = <0{\cdot}005$). This was the reverse of the authors' findings for infections by A2 influenza virus. A study was made for neutralizing antibody to adenovirus types 3 and 7, two of the serotypes most prevalent in military populations (Hilleman *et al.* 1957; McDonald *et al.* 1958), using serum specimens from 132 children aged five months to eight years. The results indicated that by the age of five to eight years neutralizing antibody for adenovirus types 3 and 7 occurred significantly more often in serum specimens from children of blood group O than in those of children of blood group A (Shedden and Potter 1964). This result indicated that the differential incidence of adenovirus infections in persons of blood groups A and O, reported by McDonald and Zuckerman (1962), may have been influenced by the higher incidence of protective antibody in blood group O persons, acquired as a result of infections in earlier years.

An examination was made for HI antibody to adenovirus types 3 and 7 in 382 serum specimens. The results for HI antibody to adenovirus type 3 are shown in Figure 3. In the three youngest age groups, aged 1–15 years, antibody was found in a significantly greater number of serum specimens from persons of blood group O, compared with those from persons of blood group A ($P = <0{\cdot}05$). However, for persons aged over 21 years, the incidence of HI antibody to adenovirus type 3 was the same for persons of blood group A and O. Thus, between the ages of 16–20 years, significantly more persons of blood group A acquire antibody than persons of blood group O (Potter 1969).

Results similar to those shown in Figure 3 were found in tests for HI antibody to adenovirus type 7 (Figure 4). In the four age groups 1–20 years, HI antibody to adenovirus type 7 was detected in significantly more persons of blood group O compared with persons of blood group A ($P = <0{\cdot}05$). In individuals aged 26–30 years or more, HI antibody to adenovirus type 7 was found with equal frequency in serum specimens from persons of blood group A and O. Thus, at the ages between 20 and 30 years, more persons of blood group A acquire antibody than those of blood group O. The results shown in Figures 3 and 4 suggest that infections by adenovirus types 3 and 7 occur in significantly more children of blood group O than those of blood group A. Later infections, at ages corresponding to that of the service population studied by McDonald and Zuckerman (1962), occur predominantly in blood group A persons as a relatively greater proportion of blood group O individuals possess protective antibody (Potter 1969).

Discussion

The results obtained in studies of antibody to adenovirus types 3 and 7 may be extended to those obtained for influenza viruses. Thus, the first wave of infections by A2 influenza virus occurred in a population totally susceptible to infection. These early infections occurred in significantly more persons of blood group O than in persons of blood group A, as shown by the study of McDonald and Zuckerman (1962), and in antibody studies of serum specimens collected in 1961–1963 (Potter and Schild 1967). Later infections would occur in a population containing a greater proportion of antibody negative persons of blood group A. In such later studies, Fakerhol *et al.* (1964) found the incidence of A2 influenza to be the same for persons of blood group A and O. In addition, the incidence of HI antibody to influenza A2 virus in serum specimens collected in 1966 was the same for persons of blood group A and O (Potter 1969). Lastly, McDonald and Zuckerman (1962) reported no difference by blood group status for infections by A1 influenza viruses in a service population during the years 1956–1962, several years after the appearance of A1 influenza viruses. No differences were found in the incidence of HI antibody to A1/FM1 influenza virus in serum specimens collected in 1966 by blood group status (Potter 1969).

The results indicate that persons of blood group O are more susceptible to infection by both A2 influenza virus and adenovirus than persons of blood group A. The relative resistance to infection present in blood group A individuals, however, is of low significance, and repeated exposure to homologous virus infection results in the initial differences being obscured. Analysis of genetic factors in determining susceptibility to virus infections should be carried out either in children, who have by reason of their age not been exposed to previous infection, or in adult populations at the time of exposure to 'new' viruses.

REFERENCES

Fakerhol, M. K., Harboe, A. and Hartmann, O. (1964). *Nature,* **203,** 1185.

Hilleman, M. R., Gauld, R. L., Butler, R. L., Stallones, R. A., Hedberg, C. L., Warfield, M. S. and Anderson, S. A. (1957). *Amer. J. Hyg.,* **66,** 29.

McDonald, J. C., Wilson, J. S., Thorburn, W. B., Holland, W. W. and Andrews, B. E. (1958). *Brit. med. J.,* **2,** 721.

McDonald, J. C. and Zuckerman, A. J. (1962). *Brit. med. J.,* **2,** 89.

Potter, C. W. and Schild, G. C. (1967). *J. Immunol.,* **98,** 1320.

Potter, C. W. (1969), *J. Hyg., Camb.,* **67,** 67.

Shedden, W. I. H. and Potter, C. W. (1964). *Nature,* **202,** 505.

Tyrrell, D. A. J., Sparrow, P. and Beare, A. S. (1968), *Nature,* **220,** 819.

Zuckerman, A. J. and McDonald, J. C. (1963). *Brit. med. J.,* **2,** 537.

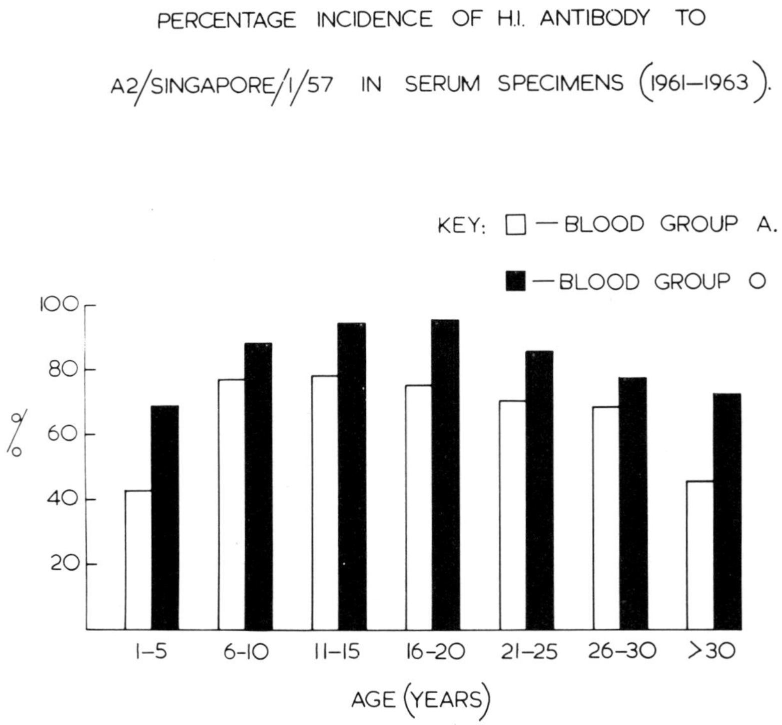

Figure 1. Reprinted by permission of the Editors, *J. Immunol.* (1967), **98,** 1320.

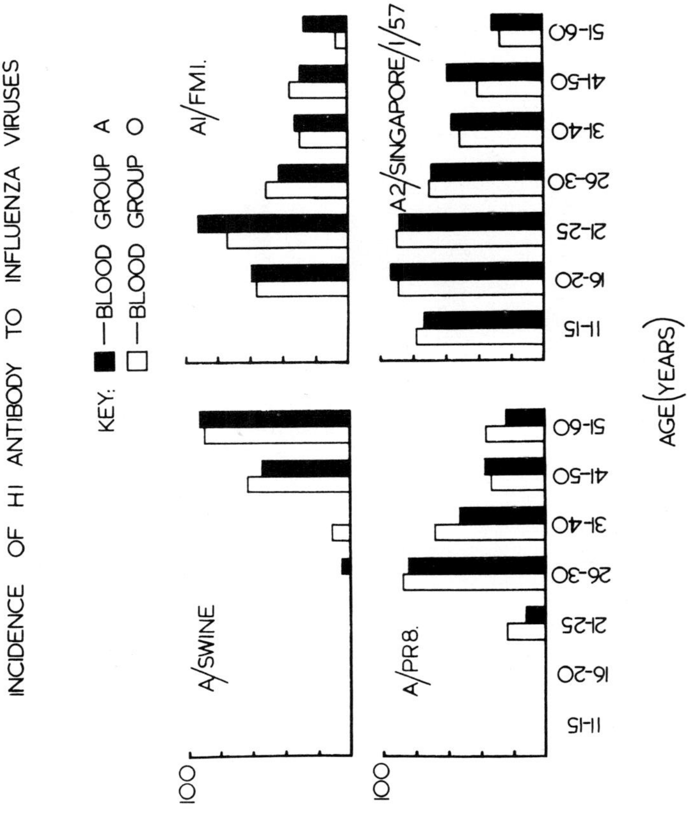

Figure 2. Reprinted by permission of the Editors, *J. Hyg., Camb.* (1969), **67,** 67.

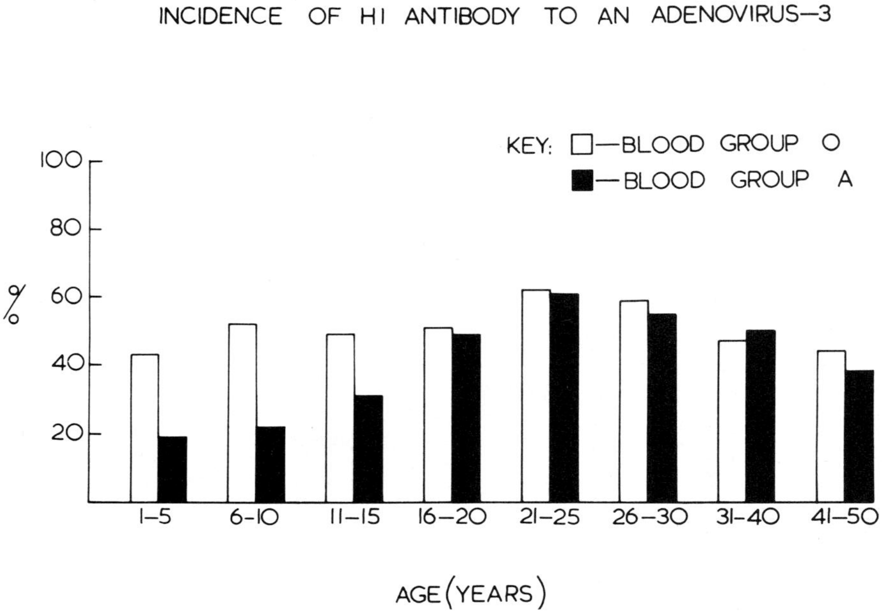

Figure 3. Reprinted by permission of the Editors, *J. Hyg., Camb.* (1969), **67,** 67.

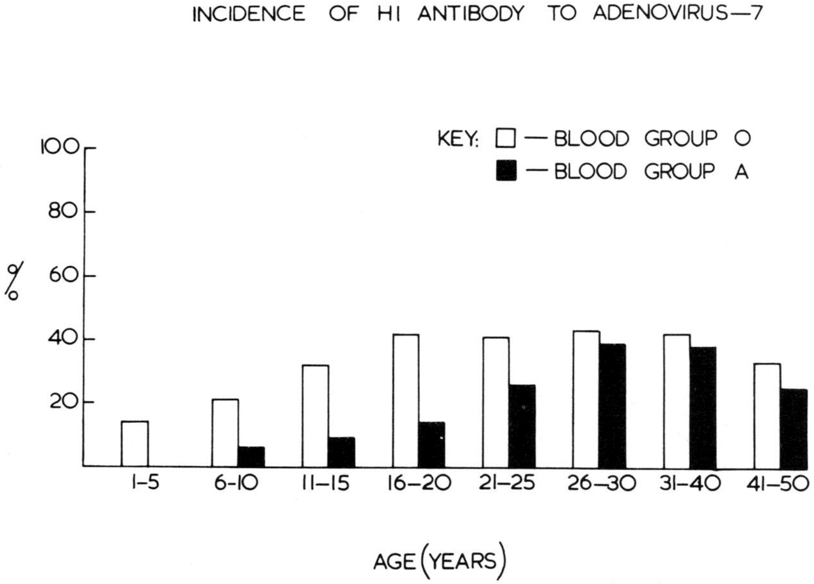

Figure 4. Reprinted by permission of the Editors, *J. Hyg., Camb* (1969), **67,** 67.

Aerosol transmission of oncogenic viruses

Gaylord E. McKissick, Laurance G. Wolfe, Robert L. Farrell, Richard A. Griesemer and A. Hellman

National Institute of Health, Bethesda, Maryland

SUMMARY

Concern about the hazards of human exposure to oncogenic viruses of animal origin is based upon the discovery that all oncogenic viruses that have been adequately studied are capable of infecting more than one species of animal. Since it was felt that the most likely route of exposure would be the aerosol route, a study was initiated to determine the feasibility of animal infection by tumor viruses via the aerosol route.

Experiments have been conducted on the aerosol transmission of four agents: Yaba virus, Rauscher Murine Leukemia virus, Adenovirus type 12, and Rous Sarcoma virus. Both Yaba and Rauscher virus are readily transmitted via the aerosol route. Evidence for transmission of the Adenovirus is presented. Evidence demonstrating transmission of the Rous Sarcoma virus to a simian species is very inconclusive.

The evidence obtained leads to the conclusion that tumor viruses can be readily transmitted via the aerosol route to susceptible hosts.

INTRODUCTION

The National Cancer Institute currently has in progress a large-scale effort in viral oncology. The conviction that there exists sufficient knowledge to study the relationship of viruses in human malignancy is based on well-founded animal studies (1). It would certainly be an oddity of nature, should man be unique and not parasitized and responsive to tumor virus transformation or oncogene expression. From a research point of view, the Special Virus Cancer Program is divided into five major efforts: Special Animal Leukemia Etiology Studies, Human Leukemia Etiology Studies, Biohazards Control and Containment, Solid Tumor Studies and Human Cancer Therapy. The biohazards working area provides the knowledge and develops techniques for preventing the spread of oncogenic viruses from animal to animal, from animal to man, from man to animal and from man to man. This concern is based on the finding that all oncogenic viruses, thus far adequately studied, are capable of infecting more than one species of animal. A major factor influencing crossing of species barriers by oncogenic viruses is the dose of the agent encountered. For example if a 10^{-2} ID_{50} of Rous Sarcoma virus is required to produce neoplasms in chicks, 10^{6} ID_{50} is required to produce neoplasms in hamsters (2). The most frequent route of exposure is most likely by inhalation. It was obviously important, therefore, to determine whether oncogenic viruses can be transmitted via the aerosol route. Representing both the DNA and RNA viruses, four animal tumor viruses were selected for study. These were: Yaba, Adeno 12, Rous Sarcoma and Rauscher Leukemia viruses. Yaba virus is infectious for tumors in hamsters. Rous Sarcoma virus is readily transmitted amongst the chicken population and has a very broad species range. Finally, the Rauscher virus serves as a model system in many tumor virology laboratories.

Aerosol transmission of tumor viruses has been reported by Burmester *et al.* (3) and Sevoian *et al.* (4) in the case of avian lymphomatosis. Wolfe *et al.* (5) in studies to be summarized in this report has produced tumors with Yaba virus transmitted via the aerosol route.

MATERIALS AND METHODS

The Rauscher Murine Leukemia virus (RMLV) used in these studies was propagated in Balb/c mice and extracted from infected spleen tissue. The extract was prepared in 0·05 M sodium citrate as a 10 per cent suspension. The total protein content of the suspension was 0·9 mg/100 ml. The titer of the preparation was $1{\cdot}0\times10^{5}$ ID_{50} ml as measured by spleen focus assay and by mouse lethality. Control animals received similar non-infective splenic material.

Weanling (5–8 weeks of age) SPF (specific pathogen free) Balb/c mice were placed in a modified Henderson apparatus (6) where the external nares and mouths of 48 mice were exposed to a Collison atomized aerosol. The animals were exposed simultaneously for 200 minutes at a relative humidity ranging between 48 and 53 per cent at a temperature of 22°C. It was calculated that this type of dynamic exposure gave a retention dose of 2675 ID_{50} effective within two years. This was based on a titre of 1×10^{5} and an average respiratory volume of 25 c.c./minute for mice (7). It was determined that 50 per cent retention of the inhaled particles of less than 1 inch diameter was achieved (8) and an air flow of 28 liters/minute utilizing 0·3 ml of viral suspension/minute was delivered. The generated aerosol was sampled in an all-glass impinger (AGI) at a flow rate of 12·5 liters/minute for 20 minutes and subsequently bioassayed.

All animals were housed in flexible isolators. Each isolator contained a total of 40 mice, four mice per cage. Twenty of these mice (two per cage) represented contact exposure animals, housed either with viral aerosol or control (non-infected splenic extract) aerosol animals. Animals were routinely observed for splenomegaly. Hemograms were performed for 40 per cent of animals developing antemortem clinical signs. The three criteria used for positive diagnosis of leukemia were

(1) splenomegaly, (2) neoplastic involvement of the spleen and (3) neoplastic cells in the liver. Three attempts to isolate virus from aerosol infected mice and re-inoculation of the splenic extract were also made.

In another study utilizing Yaba virus which induces subcutaneous histiocytomas of monkeys and man (9, 10), sixteen rhesus (Macaca Mulatta) monkeys were exposed to a dynamic aerosol of CV-1 virus-cell preparation which was scraped from the glass surface and suspended in medium 199 plus 10 per cent agamma calf serum at pH 7·2 to 7·4. The cells were ruptured by freezing and thawing four times and then centrifuged at $600 \times g$ for 10 minutes. The titer of the supernate ranged from 1×10^5 to 4×10^6 FFU/ml. The total protein concentration was 400–1000 mg./ml.

Animals were exposed to a Collison atomized aerosol in a Henderson apparatus (6). The total airflow was 12·5 liters/minute and the relative humidity ranged from 80 to 95 per cent. Fifty per cent of the particle size, as determined by Anderson sampling, ranged from 1 to 3 μ. Monkeys were exposed in groups of three for either 90 or 120 minutes. It has previously been determined that the above environmental conditions did not contribute to the loss of virus infectivity.

Periodically the aerosol was sampled in an AGI containing CV-1 cell-growth medium. This was then titered for PFU/ml of inhaled air. One animal was killed immediately after exposure and had a titer of 2×10^4 FFU of Yaba virus. This approximated the estimated retention dose.

Utilizing HEK cell propagated Adeno 12 virus, 140 newborn hamsters were exposed to a dynamic aerosol by procedures similar to those previously described at a relative humidity of 85–90 per cent. It was estimated by bioassay that animals received a 1×10^2 to 1×10^{36} $TCID_{50}$ retained lung dose. Similarly, three newborn monkeys (two rhesus, one cynomologus) and eight juvenile marmosets (cotton-top and tarmarins) were exposed to a dynamic aerosol of Rous Sarcoma virus. The relative humidity during this exposure was 90–96 per cent with an average virus titer of 6×10^4 for the marmosets and ranged between 31 and 61 per cent for the monkeys with an average virus titer of 1×10^5.

RESULTS

As is seen in Figure 1, a total of 39·5 per cent of mice exposed to RMLV developed leukemia. Eighteen mice died during the first post-exposure year and 16 during the second year. Utilizing the chi-square test and comparing the three test groups (RMLV-aerosol exposed, contact exposed and control aerosol exposed) with two degrees of freedom and the aerosol and control group compared individually with one degree of freedom, the incidence of leukemia differed significantly from the control group during the first year ($P < 0·025$). During the second year the level of significance was $P < 0·001$. Twenty-three of 34 mice developing leukemia were males. The incidence of cage mate exposed leukemia was not significant during the first year. However, during the second year it did become significant ($P < 0·001$) when compared to the control group. Of the total 14 leukemic deaths in the control exposed animals 85·7 per cent or 12 out of 14 occurred during the second year, 11 of which were males. This delayed death most likely due to the low exposure dose that the contact mice encountered and has been previously reported (11). Only 2 of 44 control mice developed leukemia. This may have been due to control aerosol exposure of splenic extract that was contaminated with intercurrent leukemia virus present in the majority of mouse colonies. In a previous pilot study, we also observed that a tenfold lower dose of RMLV was able to survive aerosolization and cause leukemia in approximately 5·0 per cent of mice exposed. All attempts to re-isolate and re-infect with splenic material from diagnosed cases of leukemia were successful.

Six of 16 animals exposed to an aerosol of Yaba virus developed histiocytomas. Only one animal developed clinical signs of the disease, as demonstrated in Figure 2. The lesions in the case were self-limiting with the tumor eventually enucleating. Evidence that the Yaba virus was the cause was confirmed by serology and electron microscopic observations for the presence of virus in the tumor. Isolation and reinoculation of the material into other monkeys was also successful. The experimental disease caused by aerosol exposure was quite different from that observed in subcutaneously induced Yaba tumors. The aerosol induced disease was extremely prolonged and widely disseminated in the lung (Figure 3). Yaba virus induced pathology was not detected in two aerosol control animals nor in any of the three contact exposed monkeys.

Of a total of 98 newborn hamsters exposed to Adeno 12 virus at 1×10^{2} $TCID_{50}$ and held for 18 months, one developed a tumor clinically attributable to the virus. This, however, could not be confirmed by serology. In a second experiment utilizing 44 animals and a virus titer of $1 \times 10^{3 \cdot 6}$ $TCID_{50}$, over 15 months after aerosolization, no neoplasms have occurred. This latter dose was sufficient to induce cellular transformation and the synthesis of Adenovirus type 12 T-antigen as evidence by the presence of T-antibody. T-antibody was present in the serum of all hamsters 4, 8 and 10 months after aerosol exposure. There is no evidence *in vivo* that normal cells can synthesize T-antigens.

Twenty months since exposure of monkeys to Rous Sarcoma virus to an aerosol, subcutaneously and by the intramuscular route, there has been demonstrated a total inability of this virus to infect the rhesus, cynomologus, marmoset and galagos. This would tend to indicate a lack of host susceptibility.

DISCUSSION

Titers of virus within the range currently utilized in many research laboratories can infect animals via the airborne route. Having therefore achieved the major objective of our study in demonstrating survival of viruses after aerosolization and infection by this route we are in a position to caution against the indiscriminate handling of such agents. Care should be taken to prevent potential human infection and even more so to prevent the confounding of experimental results. Though virus titers utilized in these studies and the exposure themselves were carried out under ideal conditions, current methods of large volume concentration of viruses do demonstrate that modern virus laboratories have the potential for creating difficulties for themselves.

REFERENCES

1. *Experimental Leukemia* (1968), Editor M. A. Rich, Appleton-Century Crafts Publishers, N.Y.
2. Bryan, W. R. (1968). The viral study of leukemia and related neoplastic diseases: a problem apart. In *Prospectives in Leukemia*, pp. 94–104. Grune and Stratton Inc., N.Y.
3. Burmester, B. R. and Gentry, R. F. A. (1954). Study of possible avenues of infection with the virus of avian visceral lymphomatosis, *Proc. Am. Vet. Med. Assoc.*, 311–317.
4. Sevoian, M., Chamberlain, D. M. and Yarose, R. H. (1963). Avian lymphomatosis. V. Air-borne transmission, *Avian Dis.*, vii, 102–105.
5. Wolfe, L. G., Griesemer, R. A. and Farrell, R. L. (1968). Experimental aerosol transmission of Yaba virus in monkeys, *J. Nat. Cancer Inst.*, **41,** 1175–1195.
6. Henderson, D. W. (1952). An apparatus for the study of airborn infection, *J. Hyg.*, **50,** 53–68.
7. Guyton, A. C. (1947). Measurement of the respiratory volume of laboratory animals, *Am. J. Physiol.*, **150,** 70–77.
8. Palm, P. E., McNermey, J. M. and Hatch, T. (1956). Respiratory dust retention in small animals, *A.M.A. Arch. Ind. Health*, **13,** 355–365.
9. Sproul, E. E., Metzgar, R. S. and Grace, J. T., Jr. (1968). The pathogenesis of Yaba virus-induced histiocytomas in primates, *J. Nat. Cancer Inst.*, **41,** 1197–1203.
10. Grace, J. T., Jr. and Mirand, E. A. (1963). Human susceptibility to a simian tumor virus, *Ann. N.Y. Acad. Sci.*, **108:** 1123–1128.
11. Rauscher, R. F. (1961). A virus-induced disease of mice characterized by erythrocytopoiesis and lymphoid leukemia, *J. Nat. Cancer Inst.*, **29,** 515–543.

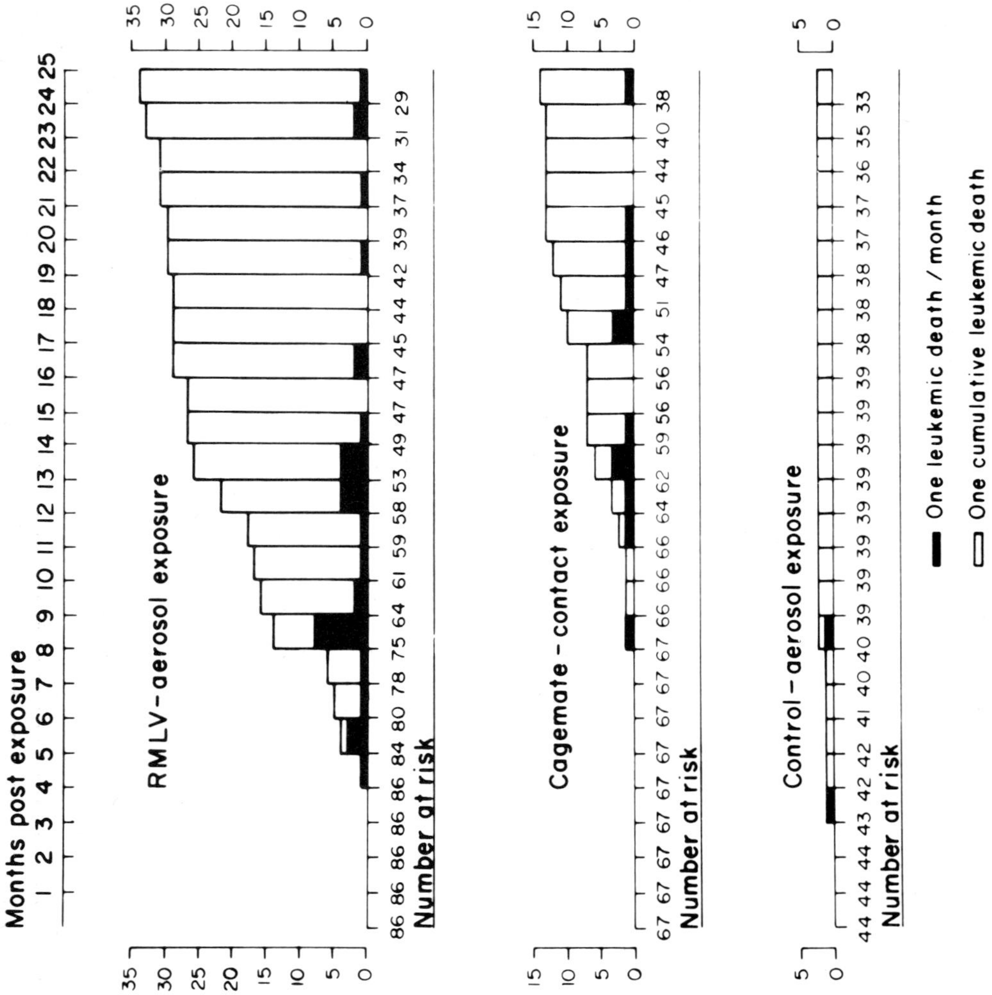
Months post exposure
1 2 3 4 5 6 7 8 9 10 11 12 13 14 15 16 17 18 19 20 21 22 23 24 25
RMLV-aerosol exposure
0 5 10 15 20 25 30 35
86 86 86 86 86 84 80 78 75 64 61 59 58 53 49 47 47 45 44 42 39 37 34 31 29
Number at risk
Cagemate - contact exposure
0 5 10 15
67 67 67 67 67 67 67 67 66 66 66 64 62 59 56 56 54 51 47 46 45 44 40 38
Number at risk
Control - aerosol exposure
0 5
44 44 44 43 42 42 41 40 40 39 39 39 39 39 39 39 38 38 38 37 37 36 35 33
Number at risk
One leukemic death / month
One cumulative leukemic death

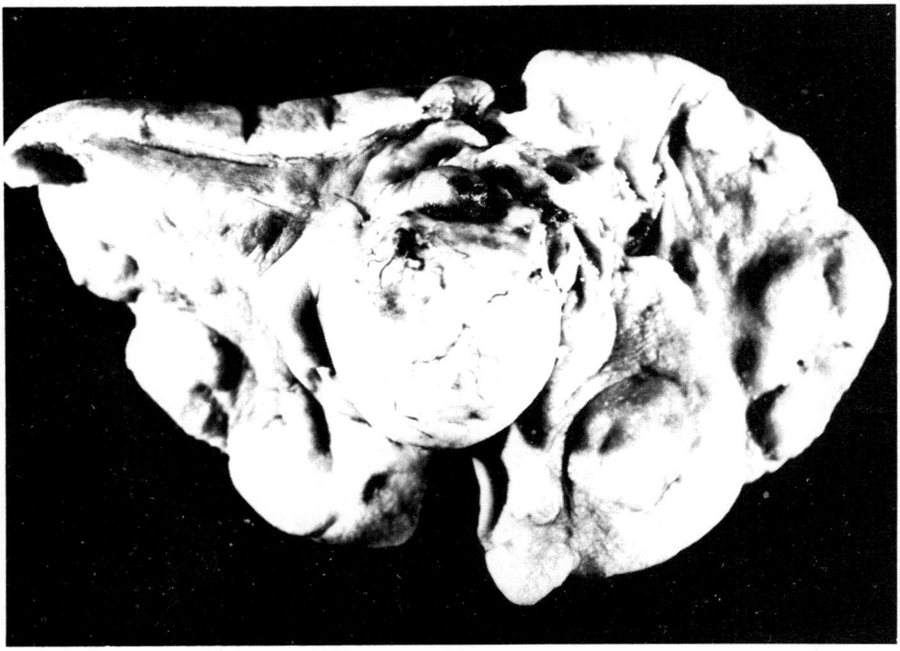

The pathogenesis of viral infections of the respiratory tract

C. A. Mims

John Curtin School of Medical Research, Australian National University, Canberra

---

The past ten years has seen little increase in our understanding of the pathogenesis of virus infections of the respiratory tract. Many new upper respiratory viruses have been identified, more is known about the antigenic variation and the epidemiology of influenza virus, but we still fail to understand the pathogenesis of interstitial pneumonia, rhinorrhea, or even pharyngitis. This brief survey will include a charting of certain areas of ignorance, most of which have already been referred to in earlier symposia, and also a number of speculations about the biology of respiratory virus infections.

Virus infections take place across body surfaces. The skin is relatively impermeable, and infection through the skin is generally associated with physical injury. This may be a scratch, as in smallpox vaccination, a large bite as in rabies infection, or a small bite as in infection with arboviruses by mosquitoes. Both the lung and alimentary canal, however, are lined by naked living cells, and are rather more readily infected. Infection of the alimentary canal has to be considered in relation to mucus, acid, bile, enzymes, anoxia and other micro-organisms, and almost nothing is known about viral infection by this route. The lung is a blind diverticulum from the alimentary canal, and as such possesses an effective cleansing mechanism for the removal of inhaled particles and other materials. These cleansing mechanisms, consisting of the alveolar macrophages and the mucociliary system of the lung and nasopharynx, deal with inhaled virus particles as they would with inert particles. It is only recently that infection has been considered in relation to these cleansing mechanisms, and more investigations of the type carried out by Dr. and Mrs. Bang (1967) are needed. If a virus particle contained in a droplet mucleus of diameter 3μ is inhaled it is likely, if it reaches an alveolus, to be phagocytosed by an alveolar macrophage. Following this, most viruses suffer the fate of inert organic particles, and are degraded and digested. This is probably as true for a rhinovirus as it would be for tobacco mosaic virus. If, however, the ingested virus replicates in the alveolar macrophage a progressive infection may have been initiated. Thus, mice inhaling an ectromelia virus aerosol show infected alveolar marcophages twelve hours later, as detected by immunofluorescence (Roberts 1962), and the infection then spreads, both in the lung and systematically. Alveolar macrophages may play a role in infections of the lower respiratory tract, especially in the case of viruses which do not readily infect respiratory epithelial cells.

If, on the other hand, an inhaled droplet lands on the mucociliary epithelium, it is likely to be treated like a particle of carbon and borne inexorably up from the lung (or down from the nasal mucosa) to the pharynx, and there swallowed. How, then, do viruses initiate infection in mucociliary epithelium? Infected epithelial cells were seen by immunofluorescent staining 10 hours after intranasal infection of mice with influenza virus (Hers *et al.* 1962), and rhinoviruses infect the nasal epithelium of man with considerable efficiency (Couch *et al.* 1966). In the first place, virus particles must pass through the mucus film and make physical contact with the cell surface, or with the cilia and microvilli which protrude from the surface. Perhaps eddies and other local movements bring particles down from the mucus surface into contact with cells. The mucus layer may be thin in parts, and there are areas where mucus flow is arrested as described by Hilding, at the tracheo-bronchial junction and larynx. The neuraminidase on the surface of influenza virus particles might theoretically help penetration of the mucus film, but evidence (Schulman *et al.* 1968) suggests that an important function of this enzyme is to liberate virus particles from the surface of infected cells. In any case, most respiratory viruses do not appear to possess enzymes, arguing against their general importance in the process of infection.

Once a virus particle has made contact with an epithelial cell, attachment to a specific receptor site may maintain this contact long enough for the next stage in infection to take place. Specific receptor sites occur in the case of influenza virus, but none have yet been described for other respiratory viruses. There is some uncertainty as to whether the next step in infection, in which the virus particle or viral genone enters the cell, necessarily involves active uptake by the cell. If uptake is an active process this implies a phagocytic act, and almost nothing is known about phagocytosis by respiratory epithelial cells. That these cells take up fluids and small molecules is indicated by the great decrease in the total volume of secretions that appears to occur by the time the mucus film reaches the trachea (Kilburn 1968). Probably larger molecules are also taken up, as suggested by the extreme potency of tetanus and botulinus toxins administered intranasally. If the response to inhaled allergens (Pepys, this symposium) are mediated by the actual entry of materials into cells, then proteins, too, may be taken up into respiratory epithelial cells. In 1930, Marion Ropes showed that the ciliated bronchial epithelial cells of rabbits were able to phagocytose carbon inhales as charcoal dust. Further experiments are needed, perhaps with the use of radioactively labelled marker particles and autoradiography, to assess the phagocytic ability of these cells. Phagocytosis by ciliated epithelial cells is not so unexpected when it is remembered that both intestinal epithelial cells and epidermal cells (Platt 1963) have been shown to have phagocytic powers.

There is one simple mechanism which may conceivably increase the efficiency of upper respiratory infection. The mucociliary flow brings particles from both the nasal cavity and lung to the pharynx, where that great muscular organ the tongue, in pushing backward with a vigorous thrust, wipes infectious material firmly against the posterior pharyngeal wall. Respiratory viruses may infect the nasal mucosa more readily than the pharynx (Buckland *et al.* 1964), but one of the earliest and most regular symptoms of upper respiratory virus infections is a sore throat, suggesting either early viral growth in this area, or alternatively an early inflammatory response in the submucous collection of lymphoid tissue in the pharynx.

The respiratory tract may be infected not only from inhaled material but also by spread of infection from elsewhere in the body. Viruses then reach the lung via the circulation, and the factors which influence their localization in alveolar or submucous blood vessels must be similar to those which have been discussed in the case of the skin (Mims 1966). As with the skin, infection may not extend far from these vessels, so that in the lower respiratory tract subsequent inflammatory and immune changes produce interstitial pneumonia. This is what happens in adult mice infected systemically with lymphocytic choriomeningitis virus (Mims, unpublished). In this case respiratory epithelial cells are not significantly infected, and only small amounts of virus are perhaps shed in alveolar cells carried up to the throat on the mucociliary tract. Alternatively the infection may spread from blood vessels to involve epithelial cells, and virus is shed into the respiratory passages with the development of bronchiolitis, bronchitis, tracheatis, etc., according to the locality. Measles virus appears to behave in this way.

The relative importance of antibodies, immune cells and interferon in controlling respiratory infections is not known. Certainly interferon could be produced by the first infected epithelial cell and thus confer resistance on nearby cells. Experiments using immunosuppressive treatments may help answer this question, and investigators still dream of the discovery of specific suppressors of interferon synthesis or of animals incapable of producing interferon.

Lesions are produced by direct viral damage to cells, and also by the inflammatory and immune host responses to infection. On the respiratory epithelium as elsewhere, viruses show a spectrum of activity, ranging from no damage to infected cells, as with certain rhinoviruses (Hoorn and Tyrrell 1969), to cell necrosis and sloughing, as is usually seen with influenza virus. As an example of an intermediate type of activity there is the noncytopathic detachment of apical viruliferous fragments of cytoplasm described by Harford and Hamelin (1952) in mice infected with influenza virus.

Infection may also lead to important changes in cell function. Ciliary beat seems a function that is not easily lost. Cilia continue to beat for several days after death (Rivera 1962) and also in shed infected and degenerating cells (Hoorn and Tyrrell 1966). There may nevertheless be gross disturbances in the orderly beat of a sheet of respiratory epithelium. In the human trachea and bronchi about one cell in five is a goblet cell (Kilburn 1968) and changes in the quantity or quality of mucus secreted by these cells may obviously be important. Goblet cells were shown to be infected when rubella was grown in human fetal trachea in culture, (Banatvala *et al.* 1969), but usually there is no information about the infection of goblet cells. Similar considerations apply with more force to the submucous seromucinous glands, whose total volume in man is 40 times that of the goblet cells (Kilburn 1968). Studies of the biochemistry and biophysics of mucus are just beginning. Changes in the secretion of surfactant might also be important in virus infections of the lung, presumably by a direct action on alveolar cells. Although a decrease in surfactant has indeed been described in pneumonic lungs (Packle 1965), this particular observation is difficult to interpret since collapse of any kind, even when produced by tying off a bronchus, leads secondarily to a decrease in surfactant.

Earlier in this symposium there was discussion of the role of immune responses in the production of lesions in the respiratory tract. Further work on the part played by immune cells, IgA and IgE immunoglobulins is needed. Immune responses appear to be important in young infants infected with respiratory syncytial virus (Chanock, this symposium) and also in those infected with measles virus after vaccination with killed measles virus vaccine (Fulginiti *et al.* 1967). For influenza virus, there is experimental evidence that immune responses produced by vaccination of mice and guinea pigs with non-infectious antigen can result in increased severity of lesions following live virus challenge (Ogasawara 1961; Tong and Fong 1964). With pox-viruses immune responses make an important contribution to the evolution of lesions in the skin (Mims 1966) and it seems likely that similar phenomena would occur in the lung. Some years ago, Klause Schell (1960) showed that C57B1 mice were especially resistant to the lethal action of ectromelia virus because of an earlier and more vigorous immune response to infection. Unexpectedly, however, they were much more susceptible than other mice when infected with large doses of virus by the respiratory route. This was attributed to their exuberant immune response which poured fluid and cells into infected lungs leading to consolidation and death.

Respiratory infections produce a variety of symptoms and signs, but their pathogenesis is generally not understood. Rhinorrhea and pharyngitis are obvious examples, but even fever is an enigma. We know that influenza virus can liberate a pyrogen from polymorphonuclear leucocytes, and that fever is produced when large amounts of killed influenza virus are injected into rabbits or man. But the mechanism of fever in natural respiratory infection is less clear and in the case of rhinoviruses, for instance, no pyrogens have been demonstrated. We have to be content with saying that the products of cell damage, viral growth, inflammatory and immune responses may generate pyrogens. Pyrogenic products absorbed from the respiratory tract are more likely to have systemic effects than after absorption from the intestine into the portal circulation, when the liver may have a detoxifying action. Doctor Pepys (this symposium) has pointed out that fever may occur as part of the reponse to respiratory allergens, so that it is not necessary to think only of pyrogens produced by viral materials.

There are certain ill-defined influences on respiratory virus infection, which are not made less important by the fact that they are difficult to investigate. First there are meterological factors. Lidwell *et al.* (1965) showed that there is often an outbreak of colds two to three days after a fall in outside temperature. Studies on the effect of humidity and other factors have not yielded clear conclusions, although experimentally, falls in relative humidity have important effects on mucociliary function (Ewart 1965). Constitutional factors are also important. People with a history of nasal, skin or gut allergy (Dowling *et al.* 1957) get many colds, and others seem never to have colds. Those with particular susceptibility to respiratory infection were in the past said to have 'second-class respiratory epithelium'. Perhaps one day such differences will be described in terms of the properties of respiratory epithelial cells, or the production of immune cells, IgA, IgE, and interferon. Psychosomatic factors, which appear to influence the occurrence of a wide variety of diseases (Hinkle and Wolff 1958), might also be expected to influence respiratory viral diseases. In certain studies, there is evidence that this is so (Cluff *et al.* 1966; Woors *et al.* 1968), but it is not always easy to distinguish between differences in the reporting of symptoms and differences in the incidence of symptoms. Stress, which interacts with para-influenza virus 3, Pasteurellas, and perhaps other agents to produce shipping fever in cattle (Thomson, this symposium), has been suggested as a factor in the high incidence of acute respiratory viral diseases in military recruits, especially those associated with Coxsackie A21 and adenovirus infections.

Respiratory viral infections are associated with crowding, or at least with certain population densities and certain types of crowding. Infection does not come from chemically polluted air, but from air polluted with other people's respiratory secretions, although the role of chemical pollution in potentiating infection and subsequent disease has been inadequately studied. Even domestic animals, when crowded together, have an increased incidence of respiratory viral infection. This is so for chickens in batteries, or for race horses at race meetings (McQueen *et al.* 1968). It has long been known that respiratory viruses tend to disappear in small isolated human communities, and in this connection it is worth comparing modern man with paleolithic man, who lived in small bands of up to fifty individuals, a hundred thousand years ago (Fenner 1968). Black (1966) has shown that a community of at least half a million is necessary if measles is to maintain itself without periodic introduction from elsewhere. One might expect that in most respiratory virus infections, even if immunity were less permanent than in the case of measles, the necessary population would be of the same order of magnitude. Chicken pox, however, can maintain itself in an isolated population as small as one thousand. This is because chicken-pox virus becomes latent in the body after initial infection, so that many years later an attack of zoster in an older individual can yield the virus necessary to infect a fresh group of susceptible children. Therefore, it is improbable that modern respiratory viruses existed in paleolithic times unless they also could become latent. Unfortunately, the cell-culture techniques that have yielded viruses from many normal human kidneys and adenoids have not been used very extensively to study tissues from the respiratory tract. Even if latency occurs after some respiratory virus infections it seems unlikely that it would be in more than a small proportion of them. At present mankind supports a total of about a hundred antigenically different respiratory virus infections, of which at least 80 affect the upper respiratory tract. This leads to the conclusion that most respiratory viruses have evolved since paleolithic times. With the great increases in human populations that occurred during the neolithic period about 10,000 years ago, and even more so with the recent astronomical increases in human numbers, there must have been great selective forces favouring the emergence of viruses that spread rapidly and could vary antigenically. They spread rapidly by their ability to induce an outpouring of virus-rich fluid and cells with accompanying sneezing in infected hosts. One might maintain that sneezing is not always a feature of upper respiratory virus infections, and that presumably infection can also be transferred by other mechanisms. One mechanism which must be considered, particularly in the transfer of viruses in large droplets over short distances, is the act of talking. The more numerous men became, the more they talked. Talking, even if less virus is shed than in a sneeze, still occurs over longer periods and could be important in the absence of sneezing. Not much is known about the factors determining the transmissibility of respiratory viruses, and further studies of the type reported by Dr. Schulman (this symposium) are needed.

The above ecological considerations help to explain the great diversity of upper respiratory viruses affecting modern man. In the coming years we are certainly going to be more crowded, and whether or not we reach the stage of living in batteries it is difficult to be optimistic about the future control of respiratory virus infection. Respiratory viruses presumably are continuing to evolve rapidly, taking advantage of our numbers and our density. Fifty years from now when polio, smallpox, measles and other viruses have been banished to the virologists' deep freeze, respiratory viruses will no doubt still be with us. Vaccines, even if effective for influenza and other viruses, seem unlikely to offer hope for the great diversity of upper respiratory viruses. Chemotherapy, if successful, would have to become continuous chemoprophylaxis. Possibly in the end the best solution will be to pay more attention to air hygiene. We condition air to keep it warm or cold, we study its bacterial purity in operating theatres and hospitals, but there must be much more that could be done in the places where we meet and exchange our viral infections. Just as the diseases spread by water were prevented by attention to the water, and in the way that diphtheria and tuberculosis were in the first place controlled by sanitary improvements, so respiratory viral diseases may eventually be controlled by improvements in air planning rather than by vaccines or by chemotherapy.

REFERENCES

Banatvala, J. E., Best, J. M. and Kistler, G. S. (1968). *Int. Symposium on Rubella Vaccines,* pp. 161–172, London (Karger, Basel/New York 1969.

Bang, B. G. and Bang, F. B. (1967). *J. Exp. Med.,* **125,** 409–427.

Black, F. L. (1966). *J. Theoret. Biol.,* **11,** 207–211.

Buckland, F. E., Bynee, M. L. and Tyrrell, D. A. J. (1964). *J. Hyg.,* **63,** 327–343.

Cluff, L. E., Canter, A. and Imboden, J. B. (1966). *Arch. Int. Med.,* **117,** 159–163.

Couch, R. B., Gate, T. R., Douglas, G., Gerone, P. J. and Knight, V. (1966). *Bact. Rev.,* **30,** 517–529.

Dowling, H. F., Jackson, G. G. and Inuye, T. (1957). *J. Lab. Clin. Med.,* **50,** 516–525.

Ewert, G. (1965). *Acta Otolaryngol.* (Suppl. 200), 1–61.

Fenner, F. (1968). *The Biology of Animal Viruses,* Vol. II, Chap. 20, Academic Press, New York and London.

Fulginiti, V. A., Eller, J. J., Downie, A. W. and Kempe, C. H. (1967). *J.A.M.A.,* **202,** 1070–1080.

Harford, C. G. and Hamelin, A. (1952). *J. Exp. Med.,* **95,** 173–190.

Hers, J. F. Ph., Mulder, J., Masurel, N., Kuip, L. V. D. and Tyrrell, D. A. J. (1962). *J. Path. Bact.,* **83,** 207–217.

Hinkle, L. E. and Wolff, H. G. (1958). *Ann. Int. Med.,* **49,** 1373–1388.

Hoorn, B. and Tyrrell, D. A. J. (1966). *Arch. ges. Virusforsch.,* **18,** 210–225.

Hoorn, B. and Tyrrell, D. A. J. (1969). *Prog. Med. Virol,* **ii,** 408–450.

Kilburn, K. H. (1968). *Am. Rev. Resp. Dis.,* **98,** 449–463.

Lidwell, O. M., Morgan, R. W. and Williams, R. E. O. (1965). *J. Hyg.,* **63,** 427–439.

McQueen, J. L., Steel, J. H. and Robinson, R. Q. (1968). *Adv. Vet. Sci.,* **12,** 285–336.

Mims, C. A. (1966). *Bact. Rev.,* **30,** 739–760.

Ogasawara, K., Aida, M. and Nagata, I. (1961). *J. Immunol.,* **86,** 599–605.

Pattle, R. E. (1965). *Physiol. Rev,* **45,** 48–79.

Platt, H. (1963). *J. Pathol. Bacteriol.,* **86,** 113–122.

Rivera, J. A. (1962). *Cilia, Ciliated Epithelium and Ciliary Activity,* Pergamon Press.

Roberts, J. A. (1962). *Brit. J. Exp. Path.,* **43,** 451–461.

Ropes, M. W. (1930). *Contrib. Embryol. Carnegie Inst.,* **22,** 77–90.

Schell, K. (1960). *Aust. J. Exp. Biol. Med. Sci.,* **38,** 271–288, 289–299.

Schulman, J. L., Khakpour, M. and Kilbourne, E. D. (1968). *J. Virol.,* **2,** 778–786.

Tong, M. J. and Fong, J. (1964). *J. Immunol.,* **93,** 35–44.

Voors, A. W., Stewart, G. T., Gutenkunst, R. R., Moldow, C. F. and Jenkins, C. D. (1968). *Am. Rev. Resp. Dis.,* **98,** 801–809.

Transmissibility as a separate genetic attribute of influenza viruses

Jerome L. Schulman

Department of Microbiology, Mount Sinai School of Medicine, New York

---

ABSTRACT

An experimental model designed to study transmission of influenza virus infection in mice was employed in an attempt to correlate differences in transmissibility of a number of strains of influenza virus and other genetically determined viral properties.

Transmissibility was found to be independent of other attributes or virulence for mice in that strains of virus less pathogenic for mice were frequently found to be more transmissible. Detailed comparison was made of two prototype strains (the poorly transmitted strain of influenza A_0/NWS and the highly transmissible Jap. 305 strain of influenza A_2). Differences in transmissibility were not related to differences in virus titres in whole lungs, but the more transmissible Jap. 305 strain was present in higher titres in bronchial secretions, and in contrast to the A_0/NWS strain could be recovered from the air surrounding infector mice. These observations suggested that differences in transmissibility might be related to neuraminidase activity and its function of influencing release of virus from cell surfaces. This theory was given further support by evidence demonstrating that antineuraminidase antibody, but not antihemagglutinin antibody, decreased the capacity of infector mice to transmit infection.

However, recombination studies in which hybrid viruses were produced possessing the hemagglutinin antigen of A_0/NWS and neuraminidase antigens derived from influenza A_2 viruses failed to provide conclusive evidence that transmissibility was transferred in association with the transfer of the genetic determinants of neuraminidase. It was concluded that transmissibility has multigenic determinants and that the antigenic nature of viral neuraminidase is only one factor which contributes to transmissibility.

Transmissibility was shown to be influenced by infectiousness. A strain of virus not previously passaged in mice was less infectious and less readily transmitted than a similar strain adapted to mice. Partial adaptation (11 serial passages) produced a strain which was more infectious and more transmissible than the unadapted virus without appreciable change in other attributes of pathogenicity for mice.

One of the absolute requirements for survival of respiratory viruses is transmissibility—a mechanism for transfer from an infected to an uninfected host where the replicative cycle can begin again. Whether transmission occurs by direct contact, or is indirect, by means of airborne dissemination of infectious virus particles, this capacity must include the ability to detach from infected cells and be incorporated in respiratory tract secretions, and the ability to survive the physical and biological stresses accompanying subsequent expulsion into the environment and infection of new hosts. To a great extent, the virus is passive in these events, depending upon the reactions of the infected host (coughing, sneezing, etc.), the nature of the environment into which the virus is introduced and the availability of new susceptibles to continue its survival. However, genetically determined viral properties also influence the extent to which virus is shed, survives in the environment and reaches new hosts.

For several years we have employed an experimental model to study transmission of influenza virus infection in mice (1, 2). The following discussion is intended to review some of the evidence we have obtained in our attempts to correlate differences in transmissibility of different strains of influenza virus with other genetically determined viral properties.

One of the earliest observations made with this model was that the transmissibility of a particular strain of influenza virus was not clearly related to other expressions of virulence for mice (1–3). As seen in Table 1, when the capacity for transmission was compared in a variety of strains of influenza virus, we found that some strains of virus, which multiplied to much higher titre in mouse lungs and produced much more extensive pneumonia, were far less readily transmitted than other strains less virulent for mice. For example, the FM-1 strain of influenza A_1 virus multiplied to much higher titre in mouse lungs and produced much more pneumonia than any of the strains of influenza A_2 virus listed on the bottom of Table 1. Nevertheless, the A_2 strains uniformly were transmitted in greater frequency.

Two prototype strains equally pathogenic for mice (the poorly transmitted NWS strain of influenza A_0 virus and the highly transmissible Jap. 305 strain of influenza A_2 virus) then were compared in an attempt to correlate the observed differences in transmissibility with other properties of the two viruses. As transmission of infection in this model was shown to be airborne (3, 4), we were particularly interested in examining aerosols of the two strains for differences in stability and for infectiousness for mice. We found no differences in the survival of artificially created aerosols of the two viruses, or in the quantity of airborne virus required to initiate infection in mice (3). However, the two strains of virus differed appreciably in the quantity of virus shed by infector mice into the environment as airborne droplet nuclei. During the period of their maximum infectiousness, virus could be recovered readily from the air surrounding mice infected with the highly transmissible Jap. 305 strain of influenza A_2 virus. In contrast, there was no recoverable airborne virus in the environment of mice infected with the poorly transmitted NWS strain (3).

These data suggested that the greater transmissibility of the Jap. 305 strain is due to some property that facilitates release from the respiratory tissues of mice and enhances the shedding of airborne virus into the environment.

Additional evidence supporting this hypothesis then was obtained from comparisons of the quantities of virus present in bronchial wash fluids following infection of mice with the same two prototype viruses. Mice were infected by aerosol with A_0/NWS or A_2/Jap. 305 virus. At intervals following infection, catheters were inserted into the tracheas of mice from each group and 1·0 ml. of sterile saline was injected and aspired according to methods described previously (5). These fluids were inoculated into fertile eggs in serial fourfold dilutions to determine the titres of infectious virus. Group suspensions of whole lungs of mice were obtained at the same intervals and were similarly inoculated into fertile eggs to measure virus titres. As illustrated in Figure 1, which summarizes four experiments, there were no appreciable differences in virus titres of ground lung suspensions of NWS and Jap. 305 infected animals during the first 72 hours of infection. In contrast, virus titres in bronchial wash fluids were appreciably higher in mice infected with Jap. 305 virus. In addition, bronchial wash fluids obtained at each interval were pooled and tested for infectivity for mice by aerosol. Serial fourfold dilutions in phosphate buffered saline were made of each pool, and groups of 10 mice each were exposed within a closed chamber to aerosols of each dilution. Two days later the number of mice infected during each exposure was determined by inoculating 10^{-2} dilutions of ground lung suspensions into eggs. For these data, it was possible to calculate the aerosol LD/50 for mice of bronchial wash fluids obtained at each interval. The results of two such experiments, summarized in the lower half of Figure 1, demonstrate that more virus, infectious for mice, was present in bronchial wash secretions of Jap. 305 infected animals. Furthermore, the highest titres of virus, infectious for mice by aerosol, were present 36 hours after infection—the time after infection when infector animals most readily transmit infection to other mice (1).

For several reasons, quantitative differences in neuraminidase activity appeared to provide the most attractive hypothesis consistent with the differences in transmissibility and the related differences in distribution of the two viruses just discussed. Appreciable evidence has been obtained suggesting that neuraminidase influences influenza virus replication by facilitating its release from cell surfaces (6–10), and enzymatic splitting of mucoproteins in bronchial secretions undoubtedly enhances the survival and mobility of influenza virus within those secretions. The influenza A_2 strains have relatively more enzymatic activity than other strains of influenza virus (11) and are uniformly more transmissible in the experimental model we have employed. Furthermore, studies of the effects of immunity on transmission of influenza virus infection in mice have provided evidence that the presence of anti-neuraminidase antibody in infector mice reduces transmission of infection whereas equal suppression of pulmonary virus replication in infector mice resulting from antihemagglutinin antibody does not affect the capacity to transmit infection (12, 13).

Accordingly, attempts were made to determine whether the property of transmissibility is transferred during genetic recombination in association with the transfer of genetic determinants of neuraminidase. The techniques by which we have prepared, isolated and identified stable recombinants of influenza A_2/R1/5 and influenza A_0/NWS viruses in our laboratory have been described previously in detail (14). In essence, the system involves double infection of eggs or human conjunctival cells with two strains of virus with identifiable genetic markers, passage of the resulting pools through appropriate antibody screens and the isolation of cloned virus by plaquing or limiting dilution. Similar methods were employed to prepare and isolate two recombinants of A_0/NWS and mouse-adapted A_0/Jap. 305 viruses. One of these possesses the hemagglutinin of A_0/NWS and neuraminidase derived from the A_2/Jap. 305 parent (A_0E). The other recombinant possesses the reciprocal pairing of surface antigens: hemagglutinin derived from A_2/Jap. 305 and neuraminidase derived from A_0NWS (A_2e). Groups of mice were infected with each of these viruses and with the two parent viruses, and transmission of infection to contact mice was determined. In addition, pulmonary virus titres were measured 48 hours after infection, and lung lesions were assessed on the seventh day of infection. As shown in Table 2, the recombinant virus having the A_2 hemagglutinin but not its enzyme (A_2e) was not as readily transmitted as its Jap. 305 (A_2E) parent, suggesting that the exchange of genetic determinants coding for A_2 enzyme (E) for determinants coding for A_0 enzyme (e) was associated with the loss of transmissibility. However, the reciprocal recombinant virus, possessing the hemagglutinin of A_0/NWS and neuraminidase derived from A_2/Jap. 305 virus (A_0E), was not appreciably better transmitted than its NWS (A_0e) parent. These results suggest that transmissibility is not entirely dependent on the same genetic factors which code for neuraminidase. In other words, the acquisition, through recombination, of A_2 neuraminidase was not sufficient in itself to endow the hybrid virus with the property of transmissibility.

Confirmation of these observations was obtained in studies of other recombinant viruses possessing the hemagglutinin of A_0/NWS and neuraminidase derived from other A_2 viruses (A_2/Taiwan/64, A_2/Hong Kong/68, A_2/Itsukaichi/65, A_2/Montevideo/67). In these experiments, the A_2 parents had not been adapted to mice previously by serial passage, and therefore were not virulent for mice and were much less readily transmitted than the mouse-adapted A_2/Jap. 305 strain. As seen in Table 3, recombinant viruses possessing A_0 hemagglutinin and neuraminidase proteins derived from A_2 virus parents were as pathogenic for mice as their A_0/NWS parent, as assessed by peak pulmonary virus titres and lung lesions, but with the exception of the recombinant which derived its enzyme from A_2/Taiwan/64, the A_0E hybrid viruses were no more transmissible. However, the relatively high frequency of transmission observed with A_0E (Taiwan) is far greater than that which has been observed in numerous experiments with the A_0/NWS parent, and is thought to represent one instance in which transmissibility was transferred in association with transfer of the genetic determinants of neuraminidase. The hypothesis we have evolved to explain these data is as follows: neuraminidase activity probably has a significant role in determining the transmissibility of influenza viruses. However, transmissibility has multigenic determinants, and also is influenced by portions of the viral genome other than those which code for the antigenic structure of neuraminidase. During recombination of A_0/NWS and A_2 viruses, only a minority of the hybrid viruses formed possess the A_2 enzyme as well as the other genetic determinants of transmissibility, and therefore few of the plaqued recombinants isolated are highly transmissible. The alternative explanation, which cannot be excluded, is that transmissibility is completely independent of neuraminidase activity, and the A_0E (Taiwan) recombinant owes its transmissibility to factors entirely unrelated to its A_2 (E) enzyme. We consider the latter to be unlikely, because of all of the earlier evidence previously discussed which links neuraminidase activity to transmissibility. In addition, it should be remembered that the genetic determinants of the antigenic structure of neuraminidase are not necessarily identical to those which determine the quantitative expression of neuraminidase activity *in vivo*. Thus many of the A_0E recombinants we have studied, although possessing enzymes antigenically identical to those of their A_2 parents, may not be as active enzymatically during viral replication in mouse lungs.

In other experiments, it was possible to demonstrate reproducibly that transmissibility could be transferred during recombination. In these experiments the parent viruses were A_0/NWS and A_1/CAM and the recombinants were identified by a stable marker present in the CAM but not the NWS parent—the formation of red bordered plaques on cultures of human conjunctival cells in the presence of neutral red in the agar overlay (15). Groups of mice were infected with each of the four viruses and peak pulmonary virus titres, lung lesions and transmission of infection were studied as in experiments described previously. Table 4, which summarizes five such experiments, demonstrates that the recombinant virus antigenically identical to its NWS parent, but possessing the capacity to form red bordered plaques (NWS r^+), was 2·5 times more transmissible than A_0/NWS virus. Conversely, the recombinant antigenically identical to CAM virus, but without its property of forming red bordered plaques (CAM r^-), was much less readily transmitted than its CAM virus parent. It should be emphasized that the genes coding for red bordered plaques are not necessarily identical to those determining transmissibility. In this instance, it is more likely that recombination, as proven by the transfer of the r^+ marker, was coincidentally associated with the transfer of genetic determinants of transmissibility.

One other factor which also affects the transmissibility of a strain of influenza virus is its infectiousness. It is evident that if a greater quantity of egg infectious particles of one strain is required to infect mice than is required for another strain, identical to the first in all other properties, the strain less infectious for mice will be transmitted less frequently. These assumptions are verified by the results of experiments shown in Table 5 in which unadapted, partially adapted and fully adapted passage lines of A_2/Jap. 305 virus were compared. The strain which was fully adapted to mice by 33 serial passages multiplied to the highest titre is mouse lungs and produced much more pneumonia than either of the other two strains. It can also be seen that partial adaptation to mice (by 11 serial passages) resulted in a marked increase in transmissibility without appreciable change in pathogenicity as indicated by peak pulmonary virus titres or the extent of pneumonia produced. Furthermore, when mice were exposed to aerosols of fourfold dilutions of each of the viruses, a greater quantity (EID/50) of unadapted virus was required to infect mice than was required for either of the other two strains. Thus an increase in infectiousness for mice following serial passage was related to an increase in transmissibility, but not to change in other indications of virulence for mice. In other experiments, we attempted to compare the transmissibility of an attenuated cold variant strain of influenza A_2/Ann Arbor/1960 virus with that of the mouse-adapted virus from which it was derived (16). High titres of the cold variant virus have been shown to infect mice intranasally and to induce antibody response without producing pneumonia or death (16). Mice were exposed to equal concentrations of the two viruses as assessed by the titres of egg infections virus in nebulizer fluids. In numerous previous experiments this concentration was calculated to deliver 1000 mouse infectious doses to each of the exposed animals. All of the infector mice exposed to the non-attenuated line were infected, and subsequently transmitted infection to 40 per cent of contact animals housed in the same cages. In contrast, none of the animals exposed to the cold variant were infected, and therefore none could transmit infection. In other words, attenuation of the parent virus by adaptation to growth at lower temperatures produced a virus which had less than 1/1000 of its original infectivity for mice. It is evident that a virus which has lost that much of its capacity to infect an animal host is no longer capable of being transmitted; even if sufficient quantities of virus could be delivered to initiate infection in the infector animal, not enough virus could be subsequently shed to infect exposed contacts.

In summary, some of the complex interrelationships which may influence the transmission of influenza virus infection are influenced by the same variety of host, environmental and viral factors which affect the initiation of infection, replication in the respiratory tract, release of virus into respiratory secretions and its survival in the environment. The viral properties which determine transmissibility are multigenic in origin and are separable from the genetic determinants of other attributes of pathogenicity such as lesion production and lethality. Among the factors directly related to transmissibility is infectiousness as expressed by the quantity of virus required to initiate infection in the respiratory tract. In addition, it is our hypothesis that the genetically determined quantitative expression of neuraminidase activity directly influences transmissibility by its effect on the release of virus into respiratory secretions and its subsequent availability for expulsion into the environment. However, the transfer of genetic determinants of neuraminidase is not associated uniformly with the transfer of transmissibility and other, unknown genetic properties also influence transmissibility.

BIBLIOGRAPHIC REFERENCES

1. Schulman, J. L. and Kilbourne, E. D. (1963). Experimental transmission of influenza virus infection in mice. I. The period of transmissibility, *J. Exp. Med.*, **118,** 257–266.
2. Schulman, J. L. (1963). The use of an animal model to study transmission of influenza virus infection, *Amer. J. Pub. Health,* **58,** 2092–2096.
3. Schulman, J. L. (1967). Experimental transmission of influenza virus infection in mice. IV. Relationship of transmissibility of different strains of virus and recovery of airborne virus in the environment of infector mice, *J. Exp. Med.*, **125,** 479–488.
4. Schulman, J. L. and Kilbourne, E. D. (1962). Airborne transmission of influenza virus infection in mice, *Nature,* **195,** 1129–1130.
5. Schulman, J. L., Khakpour, M. and Kilbourne, E. D. (1968). Protective effects of specific immunity to viral neuraminidase on influenza virus infection of mice, *J. Virol.,* **2,** 778–786.
6. Seto, J. T. and Rott, R. (1966). Functional significance of sialidase during influenza virus multiplication, *Virology,* **30,** 731–737.
7. Kilbourne, E. D., Laver, W. G., Schulman, J. L. and Webster, R. G. (1962). Antiviral activity of antiserum specific for an influenza virus neuraminidase, *J. Virol.,* **2,** 281–288.
8. Webster, R. G. and Laver, W. G. (1967). Preparation and properties of antibody specifically against the neuraminidase of influenza virus, *J. Immunol.*, **99,** 49–55.
9. Brown, J. and Laver, W. G. (1968). The effect of antineuraminidase antibody on the elution of influenza virus from cells, *J. Gen. Virol.,* **2,** 291–295.
10. Webster, R. G., Laver, W. G. and Kilbourne, E. D. (1968). Reactions of antibodies with surface antigens of influenza virus, *J. Gen. Virol.,* **3,** 315–326.
11. Seto, J. T., Hickey, B. J. and Rasmussen, A. (1959). Some biologic characteristics of Asian Influenza isolates, *Proc. Soc. Exp. Biol. Med.*, **100,** 672–676.
12. Schulman, J. L. (1967). Experimental transmission of influenza virus infection in mice. III. Differing effects of immunity induced by infection and inactivated influenza virus vaccine on transmission of infection, *J. Exp. Med.,* **125,** 467–478.
13. Schulman, J. L. (in press). Effects of immunity on transmission of influenza. Experimental studies, *Prog. Med. Virol.*
14. Kilbourne, E. D., Lief, F. S., Schulman, J. L., Jahiel, R. I. and Laver, W. G. (1967). Antigenic hybrids of influenza virus and their implications. In M. Pollard (Ed.), *Perspectives in Virology,* 5, Academic Press Inc., New York.
15. Sugiura, A. and Kilbourne, E. D. (1966). Genetic studies of influenza viruses. III. Production of plaque type recombinants with A_0 and A_1 strains, *Virology,* **29,** 84–91.
16. Maassab, H. F. (1969). Biologic and immunologic characteristics of cold adapted influenza virus, *J. Immunol.,* **102,** 728–732.

TABLE 1

Comparison of transmissibility of different strains of influenza virus

	Infector mice		
Virus	*Pulmonary virus titre, 48 hours*[a]	*Lung lesions, day 7 (%)*[b]	*no. infected*
S-15 (Swine)	7·8	45	2/20 (10)
A_0/PR8	7·5	42·5	1/20 (5)
A_0/NWS	7·6	65	3/40 (7·5)
A_0/FM-1	8·7	65	2/20 (10)
B/Lee	6·9	20	1/20 (10)
A_2/R1/5+/57[c]	6·8	2·5	6/20 (30)
A_2/Jap. 305	7·6	60	25/40 (62·5)
A_2/AA/60	7·1	20	11/20 (55)
A_2/Bethesda/63[c]	6·8	12·5	7/20 (35)
A_2/Rockville/65[c]	6·7	12·5	8/20 (40)
A_2/Aichi/68	6·9	25·0	12/20 (60)

[a] EID/50, $\log_{10}$.
[b] Extent of lung lesions (per cent).
[c] Unadapted (by serial passage) to mice.

TABLE 2

Transmissibility of recombinant viruses derived from influenza A_0/NWS and influenza A_2/Jap. 305 viruses

Virus	*Pulmonary virus titre, 48 hours*[c]	*Lung lesions, day 7 (%)*	*# Contacts infected*
A_0/NWS	7·4	55	1/30
A_2/Jap. 305	7·7	62·5	13/30
A_0E[a]	7·4	45	4/30
A_2e[b]	7·9	50	0/30

[a] E1D/50, $\log_{10}$.
[b] Hybrid virus with hemagglutinin of A_0/NWS and neuraminidase of A_2/Jap. 305 (E).
[c] Hybrid virus with hemagglutinin of A_2/Jap. 305 and neuraminidase of A_0/NWS(e).

TABLE 3

Transmissibility of recombinant viruses derived from A_0/NWS virus and four strains of influenza A_2 virus

Virus	Pulmonary virus titre, 48 hours[a]	Lung lesions, day 7 (%)	Contact mice infected	
			Number	%
A_0/NWS	7·2	70	3/60	5·0
A_2/Taiwan[b]	5·9	5	1/30	3·3
A_2/Hong Kong[b]	6·4	2·5	4/30	13·3
A_2/Itsukaichi[b]	6·0	5	4/30	13·3
A_2/Montevideo[b]	6·6	0	1/30	3·3
A_0E (Taiwan)[c]	7·6	54	12/30	40
A_0E (Hong Kong)[c]	7·7	60	2/30	6·7
A_0E (Itsukaichi)[c]	7·5	46	0/30	0
A_0E (Montevideo)[c]	8·1	45	5/30	16·7

[a] E1D/50, $\log_{10}$.
[b] Unadapted to mice (by serial passage).
[c] Recombinant viruses with hemagglutinin of NWS (A_0) and neuraminidase proteins (E) of A_2 viruses.

TABLE 4

Transmissibility of recombinant viruses derived from influenza A_0/NWS and influenza A_1/CAM viruses

Virus	Pulmonary virus titre, 48 hours[a]	Lung lesions, day 7 (%)	Contact mice infected
NWS	7·2	54	10/100
NWS r^+	7·6	50	25/100
CAM r^-	7·6	32	14/100
CAM	8·5	90	49/100

[a] E1D/50, $\log_{10}$.

TABLE 5

Effects of adaptation of influenza A_2/Jap. 305 virus to mice on virulence, infectivity and transmissibility in mice

Virus	*Mouse infectivity (MID/50)*[a]	*Pulmonary virus titre, 48 hours*[b]	*Lung lesions, day 7 (%)*	*% of contacts infected*
Jap. 305 (unadapted)	4·1	6·7	0	28
Jap. 305 (partially adapted)[c]	3·6	6·8	7·5	50
Jap. 305 (adapted)[d]	3·4	7·9	62·5	61

[a] Titre of virus in nebulizer fluid required to infect 50 per cent of mice (EID/50, $\log_{10}$).
[b] EID/50, $\log_{10}$.
[c] Eleven serial passages in mice.
[d] Thirty-three serial passages in mice.

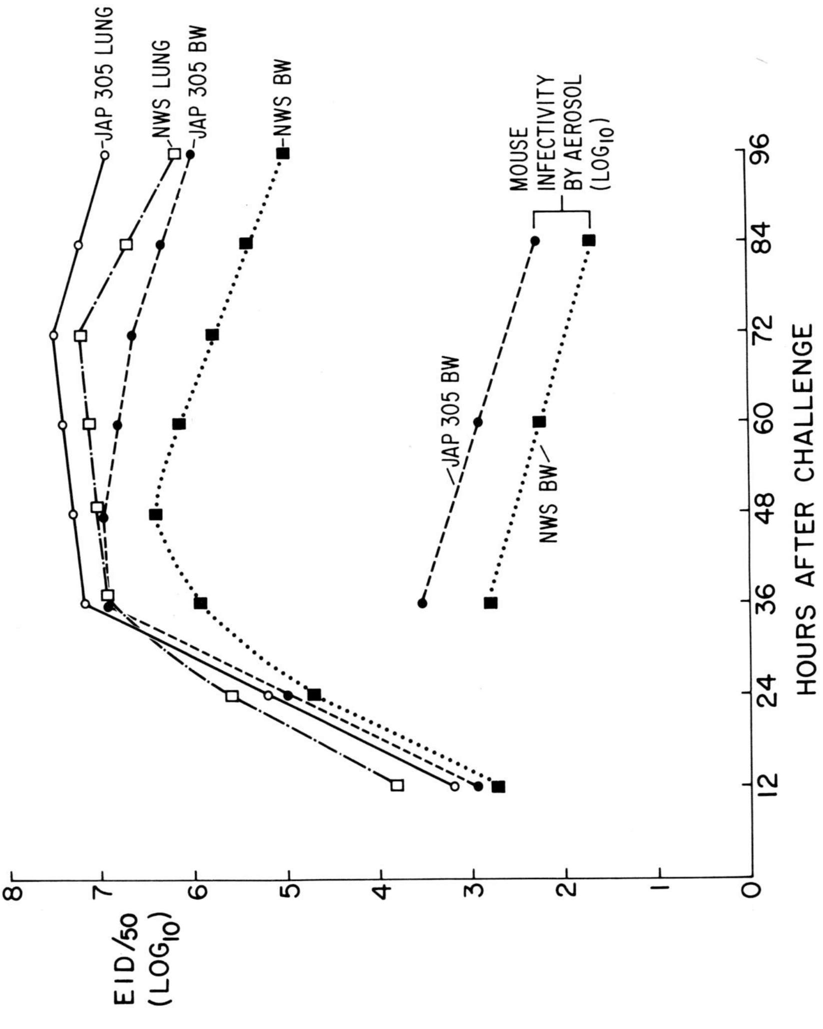

JAP 305 LUNG
NWS LUNG
JAP 305 BW
NWS BW
MOUSE INFECTIVITY BY AEROSOL (LOG10)
JAP 305 BW
NWS BW
EID/50 (LOG10)
HOURS AFTER CHALLENGE
0 1 2 3 4 5 6 7 8
12 24 36 48 60 72 84 96

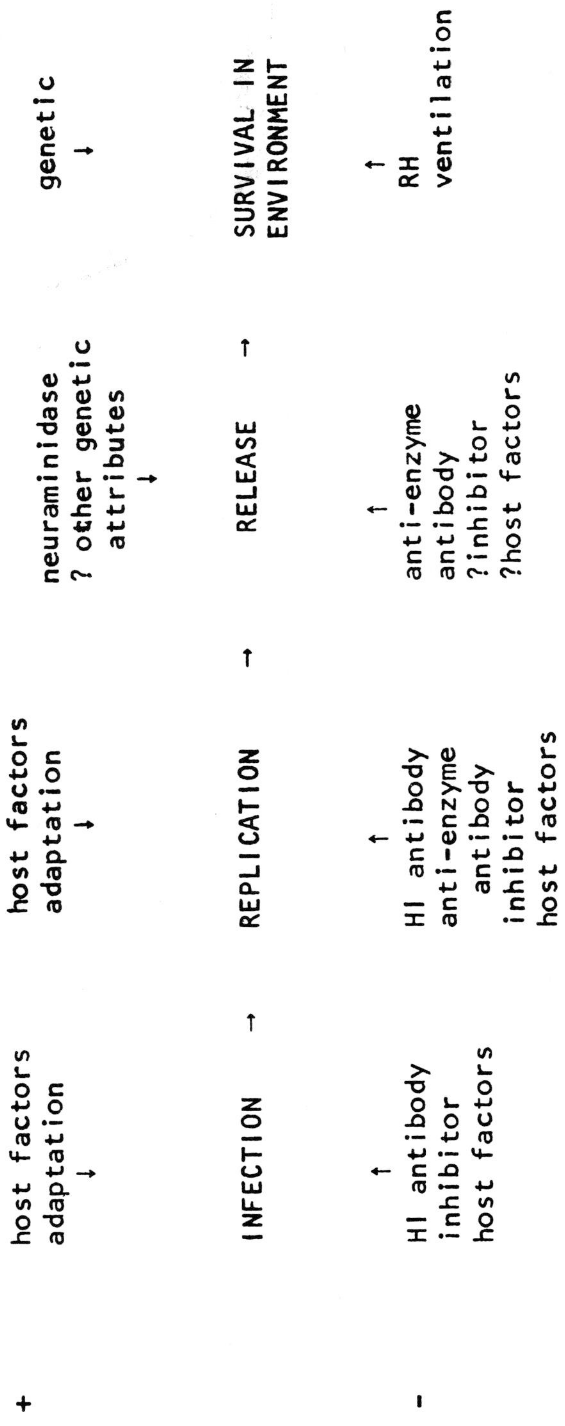
+
host factors
adaptation
host factors
adaptation
neuraminidase
? other genetic
attributes
genetic
INFECTION
REPLICATION
RELEASE
SURVIVAL IN
ENVIRONMENT
-
HI antibody
inhibitor
host factors
HI antibody
anti-enzyme
antibody
inhibitor
host factors
anti-enzyme
antibody
?inhibitor
?host factors
RH
ventilation

The inheritance of resistance to myxoviruses in A2G mice

D. Hobson

Department of Medical Microbiology, University of Liverpool

and

J. Lindenmann

Institute for Medical Microbiology, University of Zürich

The demonstration of interferon or specific antibodies in animals which survive experimental infection with virulent virus strains can usually be assumed to mean that survival is due to acquired active resistance. However, these reactions may be no more than indicators that an infecting organism has been introduced. The possibility of an innate insusceptibility of the host to infection must also be taken into account, particularly in studies involving highly inbred laboratory lines of mice, where genetic variations may be unconsciously selected.

The NWS variant of AO/WS influenza virus produced a rapidly lethal infection in most strains of mice after intracerebral inoculation of only 1–10 EID_{50} (egg infective doses) of virus. The inbred A2G line of white mice, however, survived the inoculation of 10^6–10^7 EID_{50} (Lindenmann 1962; Lindenmann, Lane and Hobson 1963). A2G mice were also insusceptible to intracerebral infections by neurotropic avian influenza viruses or by other neurotropic variants derived directly from WS or by recombination with NWS. They were, however, as highly susceptible as the Swiss white mouse line LMH to intracerebral infections by viruses other than myxoviruses, e.g. echovirus type 9, herpes simplex or arboviruses.

A similar but somewhat less solid resistance to pulmonary infection by myxoviruses was also found; the intranasal dosage of NWS required to infect 50 per cent of A2G mice was approximately 10^5-fold higher than the ID_{50} for LMH mice.

When A2G mice were cross-bred with various fully susceptible mouse lines (Lindenmann 1964; Lane 1964) resistance to myxovirus infections was conferred on the progeny, apparently as a single dominant autosomal allele. This 'resistance gene' appeared to be unique to A2G mice, since many other inbred mouse lines, even including the original strong A strain from which the A2G colony was derived, were fully susceptible to NWS infection (Lindenmann *et al.* 1963; Lindenmann and Klein 1966). The genetic peculiarity of A2G mice could possibly have been acquired by accidental illegitimate mating in the colony before it became a closed brother-sister mated herd in 1950.

The mechanisms by which genetic insusceptibility may be manifested are still not clear. Specific antibody was not induced in higher titre in A2G than in LMH mice during infection; interferon titres were no higher in A2G than in LMH mice (Lane 1964) nor was there any evidence of interference by latent virus infection. No natural inhibitors of NWS replication could be found in the serum or brain of A2G mice.

The lack of susceptibility of A2G mice is not simply due to failure of virus uptake or release. Virus adsorption-elution kinetics were similar with red cells of both A2G and LMH mice. In the intact animal, there was in fact multiplication of NWS to titres of 10^4 EID_{50}/brain in the early stages after inoculation, but the growth-rate declined sharply over the later period when virus titres were rising rapidly to final critical levels of 10^7 EID_{50}/brain in susceptible LMH mice.

An elucidation of the determinants of insusceptibility to myxovirus infection in A2G mice is of obvious academic interest, and may also provide an insight into defence mechanisms which could be employed to control the spread or severity of myxoviruses in genetically-susceptible populations.

REFERENCES

Lane, C. A. (1964). Ph.D. Thesis, University of Sheffield.

Lindenmann, J. (1962). *Proc. Soc. exp. Biol. Med.*, **116,** 506.

Lindenmann, J. and Klein, P. A. (1966). *Arch. ges. Virusforsch.*, **19,** 1.

Lindenmann, J., Lane, C. A. and Hobson, D. (1963). *J. Immunol.*, **90,** 942.

Capabilities and limitations of techniques in botanical aerobiology

J. M. Hirst

Rothamstead Experimental Station, Harpenden, Herts

I am concerned with propagules and fragments of actinomycetes, fungi, lichens, mosses, ferns and the pollen grains of coniferous and flowering plants. These particles are important to the reproduction and dispersal of lower plants; the wind pollination of flowering plants (isolation distances in plant breeding, etc.); medical and veterinary applications because of their parasitic, toxic or allergic actions; because they disperse plant parasites near and far and as fossils in geological stratigraphy (palynology).

Besides varying enormously in size, shape, structure and function, many have active discharge mechanisms controlled by wetting, drying or light, etc. Recording suction traps allow study of the liberation and concentration of spores that can be recognized visually, but none is suited to continuously depositing and growing catches on nutrient media. The Andersen and slit samplers can be useful for brief samples but not of the many spores that usually grow only on their host or parent plant species.

Particles affect man and animals most when they are inhaled; for these concentration is a valuable measurement, but the plant pathologist and palynologist are concerned mostly with the deposition of spores on plants or soils.

Despite some recent progress, much remains to be learnt about the meteorology of long-distance dispersal, the processes involved in particle deposition and how we may learn to predict deposition from measurements of spore concentration.

Recognizing airborne bacteria

W. C. Noble

St. John's Hospital for Diseases of the Skin, London

Fungal cells and pollen grains can often be identified to the species level by direct examination of debris collected from the air. This is not the case with bacterial cells; recognition of airborne bacteria normally depends on the ability of the bacteria to form visible colonies on growth media. The effects of desiccation and trauma during sampling may be to make the cells more sensitive to the composition of growth media. When the organism sought is *unlike* the majority of airborne bacteria good selective media can be designed, but there has been less success in devising a medium which will select between similar species.

With naturally generated clouds of bacteria large volume samples may be needed to recover the relatively rare pathogenic species, but the sample inevitably contains many other unwanted micro-organisms. Sampling artificial aerosols of known composition is less complicated but nevertheless differences in the quantity of the growth media become more pronounced when aged rather than fresh aerosols are sampled. Methods which depend on the use of phage or fluorescent antibody technique are simple to use on artificial aerosols but difficult to use on naturally generated clouds of bacteria. Greater attention must be paid to the design of selective media with the requirements of air sampling in mind.

Techniques for the experimental production of spore clouds of *Aspergillus fumigatus*

P. K. C. Austwick and **E. C. Appleby**
Pathology Department, Royal Veterinary College, London

The experimental infection of animals with fungal spores was first attempted 100 years ago but few quantitative determinations of the effect of inhaled spores have been made since. During a period of ten years a search has been made for reliable techniques which will give constant and reproducible aerosol clouds of *Aspergillus fumigatus* spores for this purpose and the search is not yet complete.

A detailed account of the methods used and the results of experiments involving heifers, piglets, adult fowls, adult ducks and young chicks is in course of preparation, but it was felt that a brief account of some of the technical problems involved might be of interest to this meeting. These can be summarized under the headings: spore production, aerosol generation, aerosol sampling and animal exposure.

Spore production. The fungus is grown on malt agar in Roux flasks and incubated at 37°C. The yield of spores almost doubles each 24 hours up to 60 hours and 0·25 gram can be produced by each flask. Approximately 700 grams have so far been produced in a total of 54 batches. The spores are harvested by washing off, drying, sieving and grinding, and this is done in an isolation area. Protective clothing including a respirator with a filter of penetration value of not more than 0·008 per cent is worn.

Aerosol generation. Aerosol production was first attempted from spore suspensions using a Collison Spray, then from a compacted mass of spores in the Wright Dust Feed Mechanism (B. M. Wright 1950) and finally with loose spores in a modified Perkins Particle Dispenser (W. A. Perkins *et al.* 1952). Only the last apparatus gave any measure of control over the airborne concentrations produced. Cloud levels of 135×10^6 and 41×10^6 were maintained in the experiments, but with lower levels required for the exposure of young chicks, the low speed of the spore delivery wheel allowed the spores to accumulate at the sides of the wheel and bring it to a halt. This difficulty has not yet been overcome.

Aerosol sampling. All glass Porton-type impingers and millipore filters have been used for sampling. The millipore filters are carried in a three-part plastic unit which forms a convenient holder for insertion into the exposure chamber and for weighing the sample. Careful attention to desiccation is necessary to achieve accurate results.

Animal exposure. The Henderson apparatus (D. W. Henderson 1952) was used in the first experiments. Later open-ended apparatus were designed. A large one having an airflow of 1000 litres/minute was used for the exposure of heifers, piglets and adult ducks and chickens. A smaller apparatus with an airflow of 100 litres/minute, housed in a specially adapted loose box, was used for young chicks.

Sampling of airborne viruses

G. Thomas

Microbiological Research Establishment, Porton

Many samplers have been used for collecting airborne viruses, e.g. sedimentation plates, air centrifuges, impingers, bubblers, filters, slit samplers, electrostatic precipitators, cyclones, etc. Under laboratory conditions samplers such as impingers are satisfactory but natural virus aerosols have proved difficult to sample effectively with these techniques. Successful collection has only been achieved by sampling large volumes of air. Most samplers in common use (10–15 litres/minute) are unable to operate for long periods without serious evaporative losses. Large-volume samplers (500–10,000 litres/minute) have successfully sampled natural aerosols of coxsackie virus A21, adenovirus 4 and rabies virus where low-volume samplers such as impingers had failed. Such samplers are, however, expensive and rather complicated in design and operation and are not yet commonly in use.

The extension of the sampling capabilities of the standard slit sampler (Casella), enabling it to sample for up to one hour (1700 litres) by devising a special adhesive surface on which to impact airborne particles, has been successful with natural aerosols of rabbit pox and human smallpox. This technique has also proved applicable to the Andersen sampler so that particle-size range can be established.

Air sampling for allergens

R. R. Davies

Wright Fleming Institute of Microbiology, London

The basic principles of aerosol sampling by impaction and impingement are common to many disciplines including allergy. Since allergens range from large pollen grains and clumps of algal cells, which are greater than 50 microns in diameter, to spores which may be smaller than 2 microns in diameter and which may or may not be viable, or only capable of growth on a particular medium, and even have more than one growth form, emphasis will be given to the variations in technique required.

With the changing seasons of the year, the air outdoors contains a succession of spore clouds responsible for seasonal disease. The continuous monitoring of such clouds over urban areas as an index of patient exposure and the factors affecting choice of a sampling site will be discussed. Consideration will be given to the limitations imposed by background contamination, the use of 'spot sampling' in comparison with continuous sampling and the identification of allergenic dust which cannot be identified by visual or cultural methods.

The Naval Biological Laboratory programmed environmental aerosol facility: theory and operation

Leonard J. Goldberg
Naval Biological Laboratory, California

Early studies on viable aerosol survival of micro-organisms were usually conducted in static chambers fitted with fans to minimize settling. To provide longer observation times larger chambers had to be constructed since mean aerosol holding time is directly proportional to the effective height of the holding chamber.

An alternative concept for long-term aerosol holding led to my development of the slowly rotating drum. Mathematical considerations of the behaviour of aerosol particles in a rotating drum will be summarized. Additionally, some of the major technical aspects of the operation of this facility will be presented; these include air handling, aerosol production, aerosol monitoring, animal exposure and viable aerosol sampling. Data obtained on the aerosol persistence of selected hospital pathogens will be presented to illustrate the environmental range of the programmed environmental aerosol facility.

The recovery of airborne respiratory syncytial virus

J. Rechsteiner

Rijks Instituut voor de Volksgezondheid, Utrecht

One millilitre of a partially purified suspension of respiratory syncytial (RS) virus, containing about 10^7 p.f.u./ml., was aerosolized into a static system containing 2000 litres of air. By means of a FK8 direct type atomizer from this suspension about 10^{10} droplets with a mean diameter of 4·9 μ were generated. The cloud was kept at 20·5° at relative humidities (RH) ranging from 20 per cent to 90 per cent.

When the aerosol was sampled with an all-glass capillary impinger very low recoveries were found, apparently by inactivation due to shearing forces during sampling. The use of the multistage liquid impinger (May 1966) resulted in very satisfactory recoveries, especially in the bottom compartment. With this compartment alone the survival of RS virus in air could be studied. Recoveries at one minute after aerosolization were high (81 per cent) at high relative humidities and low at low RH, with a minimal value of 5 per cent at 40 per cent RH. Though, in part, this influencing of RH might be due to decreasing collection efficiency because of smaller particles at lower RH, the survival curves with time showed that inactivation in air also contributed to the low recoveries.

Techniques in aerobiology

Résumé: presented by **J. D. Morton**

Dr. Appleby discussed the production of clouds of *Aspergillus fumigatus* spores for experimental exposure of a range of animals. Spraying of aqueous suspensions from a Collison atomizer was unsatisfactory, as was the use of the Wright Dust Feed Mechanism. Moderate success was obtained with the Perkins (*scil.* 'Stanford' or 'Metronics') Dispenser, but there are unsolved problems in controlled generation of the lower concentrations.

The range of particle sizes involved in allergy, and seasonal variations in form and viability within species, aggravate problems in sampling and recognition which are already considerable in the presence of the usual large and variable background. *Dr. Davies* described the use of instruments such as the Casella slit sampler and Hirst spore trap, expressing preference for the latter. He showed correlations of allergen count and patient response which indicated seasonal development of some immunity.

Dr. Thomas has explored a range of samplers for collection of virus aerosols under natural conditions. This is much more difficult than sampling laboratory aerosols and success hitherto has been limited almost entirely to complex, expensive 'large-volume' samplers. He had developed an adhesive surface which raised the capability of the standard Casella slit sampler and Anderson sampler to much longer periods of aspiration. A notable success was identification of airborne smallpox virus in a hospital ward.

Dr. Hirst emphasized the importance of visual recognition in identifying fungal spores, pollen grains, etc., which cannot be grown except on their natural 'target'. He also discussed the importance of natural deposition as opposed to the concentration of the passing cloud, which governs respiratory dosage. He reported interesting results on the trajectory and vertical concentration profiles of particles over small and large distances of travel.

The recognition of airborne bacteria is difficult mainly for two reasons: they are visually unrecognizable except by characteristics of growth, and pathogens are normally overwhelmed by other species and miscellaneous background. *Dr. Noble* discussed the limitations of phage and fluorescent antibody typing of natural aerosols. Selective media are more useful, but they also have deficiencies and there is considerable scope for development. He mentioned that moribund organisms are more fastidious than vigorous organisms in their nutrient requirements.

Dr. Rechsteiner had found the capillary impinger unsatisfactory for sampling respiratory syncytial virus, although it was suitable in similar circumstances for a variety of other viruses. May's multistage liquid impinger gave good results and Dr. Rechsteiner has used the final stage alone (as a subsonic impinger) with success in aerosol survival experiments.

Dr. Goldberg described features of the Naval Biological Laboratory aerosol facility. He discussed the theory of the rotating drum for suspending aerosols, and showed the system for humidity control and aerosol transport. The forward light scattering meter for monitoring aerosol concentration was described. Some typical results emphasized the very large range of decay rates that can be easily handled by the overall system.

My own feeling about advances in technique since the first symposium six years ago is that they are mostly details of unspectacular consolidation rather than notable studies forward. The most interesting ones referred to in this seminar, and new since 1963, were Thomas's success in sampling natural virus aerosols, Hirst's long-range diffusion studies, May's multistage liquid impinger, and Druett and May's microthread technique. I feel that the greatest need is for aerosol technology in all phases of cloud generation, cloud holding, sampling and assessment, that is much more closely related to the processes of airborne infection, which was the subject of our symposium.

Lung defence to infectious agents

Ragnar Rylander

Department of Environmental Health, Karolinska Institute, Stockholm, Sweden

The basic defence mechanisms of the lung to infectious agents—the mechanical removal and the biocidal activity of the lung—are reviewed. The principal changes in their function after exposure to environmental irritants are discussed, with special emphasis on the importance of these changes versus the occurrence of infectious disease.

A method to measure the mechanical removal of bacteria and the bactericidal effect of the lung in the same experimental animal is described and results from exposure experiments are discussed.

Exposure to various environmental agents resulted in changes in the mechanical mucus transport and/or the efficiency of the bactericidal mechanisms. Although the changes found were relatively small, they are considered to be of importance for the development of disease after inhalation of infectious agents especially in cases where the reserve capacity of the lung defence to infection is reduced. The initial change found could also predispose for aggravated effects of a second or a third exposure to environmental agents whereby a subsequent exposure to infectious agents could cause disease.

The importance of particle size in airborne viral respiratory infections

V. Knight

Baylor College of Medicine, Houston, Texas

Since particle size determines the physical stability of aerosols and, when inhaled, the sites of deposition and the degree of retention, knowledge of the size of the particles of aerosols which may transmit human virus infections is of immense interest.

Our work has been chiefly concerned with coxsackie virus A type 21 (A-21), but, in addition, infective doses by aerosol were measured for adenovirus type 4, strains of influenza virus A_2 and rhinovirus. The findings are summarized below.

Sneezes and coughs and other echalatory activities stimulated in normal subjects produce heterogeneous aerosols with large numbers of small particles and smaller numbers of large particles.

A-21 was demonstrated in the bolus of air of sneezes and coughs of infected volunteers under conditions consistent with its presence in both the small and large particles of the discharge.

A-21 was present in the air of rooms occupied by infected volunteers in amounts sufficient, if inhaled, to cause infection in a susceptible person. It appeared that particles large enough to settle rapidly would not be recovered by the sampling procedure used.

Small doses of respiratory viruses given by small particle aerosol are infective. These studies were based on determinations as 50 per cent tissue-culture infectious doses ($TCID_{50}$) and by electron microscopic counts of viral particulates in the inoculum.

We also found differences in infective doses of different viral agents when administered in nasal drops or in small particle aerosol. Such differences might be related to patterns of natural spread of these infections and emphasize the danger of attempting to generalize about mechanisms of airborne transmission of viral infection.

Finally, in a study of volunteers with A-21, we obtained definite evidence of airborne transmission that is most consistent with transmission by small particle aerosol.

Pathogenesis of pneumonia

R. G. Thomson

University of Guelph, Ontario

When cattle are stressed by transport for long distances they often develop a highly fatal fibrinous pneumonia. Usually pure cultures of *Pasteurella hemolytica* are grown from the lesion although there is considerable speculation that parainfluenza-3 (PI-3) virus is instrumental early in the disease. We are studying the pathogenesis of this pneumonic lesion.

Any investigation of this disease is complicated by two factors. First, clinicians believe that stress is the most important inducing factor and this is supported by the fact that a history of recent movement of cattle has always been associated with the disease. Second, virologists believe that parainfluenza-3 (PI-3) virus is the inducing factor because infection with this virus often occurs when cattle are brought together and moved, not unlike military recruits. The possible interaction of bacteria and virus in this disease has greatly stimulated our interest. The common assumption that pneumonias are viral in origin and only secondarily complicated by bacteria may not be true.

Marked changes occur in the nasal bacterial flora when the cattle are stressed. *Past. hemolytica* increases in frequency of isolation and in numbers of organisms present in the nasal flora and these changes seem to predispose to entry into the lung. The immune status of the host appears to be significant in determining the development of disease. Although infection with PI-3 virus occurs in the animals it does not appear to be decisive in determining whether or not pneumonia develops.

The investigation of the natural disease has demonstrated that *Past. hemolytica* is more significant than PI-3 virus in causing clinical illness leading to death in the naturally occurring disease. However, if the early stages of the disease are considered, the evidence concerning etiology is still not clear. Although we have tended to somewhat discredit PI-3 virus as an etiological factor, it could have a role in the early stages. In the animals that were killed immediately after qualifying as sick, pneumonic lesions were present and *Past. hemolytica* was present but not in large numbers.

Although we are working with an economically important disease of cattle, it has many aspects that apply directly to the pathogenesis of respiratory infections in any species. In particular, the possible interrelationship of bacteria and virus in inducing pneumonia has broad application. Also, the role of nasal flora, especially the numbers of organisms present, has not been adequately assessed in the pathogenesis of respiratory infections. The role of nasal antibody seems to be more clear in viral infections than bacterial infections. It has been suggested that nasal antibody is of greater significance in protection from some respiratory infections than is humoral antibody. The role of the two immunologic systems in prevention of the disease state, in particular a severe state of disease involving bacteria, has not been assessed. Our observation on animals clinically ill and on fatal cases suggest that animals which produce antibody rapidly probably live, but in those in which the disease outrushes antibody production, death is likely to follow. Under natural conditions the stress of assembling large numbers of animals and moving them long distances is probably highly significant and must be evaluated along with the microbiological and immunological data.

Allergic pulmonary disease due to inhaled organic dusts

J. Pepys

Department of Clinical Immunology, Insitute of Diseases of the Chest, Brompton Hospital, London

---

The main factors influencing the production of different allergic pulmonary diseases by inhaled organic dusts are:

1. The immunological reactivity of the subject.
2. The nature of the inhaled dust. For example, certain fungal spores (*Aspergillus fumigatus*) can grow in the respiratory tract, others act only as sources of antigen. The particle size of the inhaled dust determines its penetration into, and retention in, the lungs.
3. The circumstances of exposure, whether regular or intermittent, heavy or light.

A very wide range of organic dusts of vegetable, animal, fungal and other origin are known to cause allergic pulmonary disease.

How do airborne particles initiate infection?

Résumé: presented by **D. A. J. Tyrrell**

Dr. Rylander showed by his technique that although SO_2 gas alone had no measurable effect, SO_2 and coal dust or cigarette smoke inhibited the mechanical clearance of organisms and their removal by phagocytes. Others had shown that influenza virus infection reduced the clearance of bacteria, the maximum effect being at 5–6 days after infection and possibly by an effect both on ciliated cells and lung phagocytes. Alcohol apparently has different effects at different doses: low concentrations might increase clearance by phagocytes, but if pathogens such as pneumococci were used the bacteria could multiply and possibly damage leucocytes.

Dr. Knight's results suggested that adenovirus type 4 infected the lower respiratory tract more readily than the upper, influenza infected both equally easily, and coxsackie virus A21 and rhinovirus infected upper respiratory tract more readily than the lower. Less than one tissue culture dose of rhinovirus was sufficient to infect the human nose and also to infect an organ culture of human nasal epithelium; these organ cultures also yielded more virus when infected than did tracheal epithelium. Theoretical studies by Landahl had shown that in space vehicles the absence of gravity would reduce the amount of fine droplets retained in the lungs, and increase the amount of large drops persisting in the air and likely to be trapped in the upper respiratory tract.

A question was asked about the use of filters and ventilation to prevent virus infections in hospitals. Although a filter which held back 1 μ particles would probably hold back most virus particles, even a 5 μ filter might hold back a great many. The trouble was that no one knew what size particles were really carrying virus in natural situations, and though whole small particles might be important in transmission of influenza or adenoviruses, they were possibly not important in rhinovirus infections. Ordinary turbulent ventilation would be rather inefficient at preventing infection, and even rapid air changes might have little effect if air was trapped by ward furniture, or transmission was probably by inhalation of a cloud just produced by a cough or sneeze. If susceptibility was very variable or the virus very aero-stable then even substantial reductions in numbers of particles might have little effect on the frequency of transmission.

Following *Dr. Thomson's* paper it was mentioned that small droplets might be seen travelling *down* the trachea in animal experiments, and that in this way nasal bacteria might be carried down the respiratory tract, apparently against the stream of mucus. It seemed that bacteria were responsible for most of the disease manifestations of pasteurellosis in cattle since antibiotics prevented it completely, provided they were given in time. The parainfluenza virus infections might be necessary for pulmonary disease, just as influenza might pave the way for bacterial pneumonia in man, but antiviral vaccines were of little value, and cases occurred in the absence of virus infection.

In the U.K. the disease pattern is rather different. It occurs in calves a few weeks old and 80 per cent of outbreaks seem to be due to parainfluenza 3 virus, and adenovirus or a bedsonia infection.

Dr. Pepys pointed out that 10 per cent of individuals were atopic, and readily developed a type 1 hypersensitivity in which reagin and antigen reacted and caused an immediate reaction both in skin and lung. Others were non-atopic but could be sensitized by prolonged exposure to antigens when they developed precipitins which combined with antigens and fixed complement, thus leading to the delayed production of oedematous infiltrated lesions of the Arthus type. If such antigens were inhaled into the lung there was most characteristically an interstitial pneumonia. Very many types of spores, bird proteins, etc., gave rise to this extrinsic allergic alveolitis which might be confused clinically with infections. Such people might have a delayed type of asthma. Furthermore, long-term response to small amounts of such antigens could lead to chronic fibrosis of the lung.

In the bronchi, antigen was probably dissolved from the spores and entered the mucosa where it might combine with antibody and complement. During infections with bacteria there was no evidence of such reactions, for example, in chronic bronchitic patients with circulating antibody against *Haemophilus*. The reason was not known. Allergic subjects might have profuse nasal discharge during colds, but it was not known whether they were allergic to viral proteins.

Dr. Tyrrell showed electron-micrographs of organ cultures of ciliated epithelium taken by Dr. R. Dourmashkin. These had been exposed to high concentrations of purified virus and showed vaccinia virus near to the cilia but apparently held above them in a layer of mucus. Influenza virus particles, on the other hand, were dispersed among the cilia and many were clearly absorbed to the membrane of the cilia, and occasionally of the microvilli. It was believed that this represented the first step of the infection of ciliated cells by the virus.